The Coding Symbols Handbook

A Beginner's Visual Guide

By

Susie Hala

ISBN 979-8-9991963-3-0
First Edition: October 7, 2025
Published by Hala Publishing
Downey, California, USA

For permissions requests, bulk orders, or other inquiries, contact:
halaprosper1@gmail.com

Cover design by Susie Hala
Interior formatting by Susie Hala

Trademarks:
All programming language names, software products, and company names mentioned in this book are trademarks or registered trademarks of their respective owners. The author is not affiliated with, endorsed by, or sponsored by any of the companies or organizations mentioned in this book.

Disclaimer and Limitation of Liability

Educational Purpose Only

This book is designed to provide educational information about programming symbols and syntax. It is sold and distributed with the understanding that the author is not engaged in rendering professional programming services, software engineering advice, or computer science instruction. The author is a fellow learner sharing knowledge gained through personal study and experience.

No Professional Advice

The content of this book is for general informational and educational purposes only. It should not be considered professional programming advice, a computer science curriculum, or a substitute for formal education. Readers should verify all code examples and syntax rules with official documentation and should not rely solely on this book for production code or professional projects.

Accuracy and Updates

While every effort has been made to ensure accuracy, programming languages evolve and syntax rules may change over time. The author makes no warranties, express or implied, about the completeness, accuracy, or reliability of the content. Code examples are provided for learning and may not work identically in all environments or future versions of programming languages.

No Warranty & Limitation of Liability

This book is provided "as is," without warranty of any kind. The author shall not be liable for any damages arising from its use, including but not limited to data loss, project failure, or lost productivity.

Third-Party Resources

Any third-party websites, tools, or resources mentioned are provided for convenience only. The author is not responsible for their availability, content, or changes after publication.

Personal Responsibility

Learning programming requires practice and verification. Individual results will vary. Readers are responsible for testing code, verifying information, and choosing appropriate learning resources.

Feedback and Corrections

The author welcomes suggestions and corrections at: halaprosper1@gmail.com.

By using this book, you acknowledge that you have read and accepted this disclaimer and its terms.

Table of Contents

The Symbol Families — How Programming Symbols Connect and Work Together

Programming is like learning a new written language.
Just as letters form words and punctuation shapes meaning, **symbols** give code its structure, logic, and power.
To make them easier to remember, we've grouped them into **families** — sets of symbols that share a common purpose.

1. Grouping & Structure Family

These symbols **organize code** and define where data and logic belong.

Symbol	Purpose	Example
()	Group expressions, call functions	`result = (2 + 3) * 4`
[]	Create or access lists/arrays	`items = [1, 2, 3]`
{}	Define blocks, objects, or dictionaries	`let user = { name: "Alice" };`
<>	Used in templates/generics (C++, Java)	`List<Integer> numbers`

```
numbers = [1, 2, 3]
if (2 + 3) * 4 > 10:
    print("Big number!")
```

2. Math & Arithmetic Family

Symbols that perform **calculations** and transform numbers.

Symbol	Meaning	Example
+ - * /	Add, subtract, multiply, divide	`total = price * quantity`
%	Modulo (remainder)	`10 % 3 # 1`
//	Floor division	`10 // 3 # 3`
**	Exponentiation	`2 ** 8 # 256`
+= -= *= /=	Compound shortcuts	`score += 5`

```
total = 19.99 * 3
remainder = 10 % 3        # 1
power = 2 ** 5            # 32
```

6

3. Logic & Boolean Family

These symbols **connect conditions** and control decision making.

Symbol	Purpose	Example
and / &&	True only if both are true	if age >= 18 and has_license:
`or /		`
not / !	Invert truth	if not logged_in:
== != > < >= <=	Compare values	if score != 100:

```
if (age >= 18 && hasLicense) {
    console.log("You can drive");
}
```

4. Bitwise & Binary Family

Symbols that work **at the raw 1s and 0s level** for speed and control.

Symbol	Purpose	Example
&	Bitwise AND	5 & 3 # 1
`	`	Bitwise OR
^	XOR (different bits)	5 ^ 3 # 6
~	NOT (flip bits)	~5 # -6
<< >>	Shift bits left/right	8 << 2 # 32

```
value = 5 & 3        # 1
shifted = 8 << 1     # 16
```

5. Assignment Family

Symbols that **store or update values**.

Symbol	Purpose	Example
=	Assign value	x = 10
+= -= *= /= //= %= **=	Update with operation	count += 5

```
count = 10
count += 3     # now 13
```

6. Access & Reference Family

Used to **reach inside objects or namespaces**.

Symbol	Purpose	Example
.	Access object property or method	user.name
->	Pointer/member access (C/C++)	ptr->value

Symbol	Purpose	Example
::	Scope resolution (C++, Java)	`Math::pow`

```
let user = { name: "Alice" };
console.log(user.name);
```

7. Comment & Documentation Family

Symbols for **notes and explanations** in code.

Symbol	Purpose	Example
#	Single-line comment (Python)	`# This explains the code`
//	Single-line comment (JS/Java)	`// Log user name`
/* */	Multi-line comment	`/* Block of notes */`

```
# Calculate total price
total = price * quantity
```

8. Special & Language-Specific Family

Unique helpers and shortcuts.

Symbol	Purpose	Example
;	End statements (Java, JS)	`int x = 5;`
:	Dictionaries, type hints, labels	`data = {"key": "value"}`
@	Decorators/annotations	`@dataclass`
$	Template variables (JS, PHP)	`` `Hello ${name}` ``
? :	Ternary conditional	`let msg = loggedIn ? "Hi" : "Please log in";`

Troubleshooting Common Symbol Mistakes

1. **Mismatched brackets** → Always close `()` `[]` `{}` correctly.
2. **Using = instead of ==** → One assigns, the other compares.
3. **Mixing & with &&** → Bitwise vs logical confusion.
4. **Forgetting** `;` in languages that require it.
5. **Assuming symbols behave the same in all languages** (e.g., `**` works in Python but not Java).

Visual Cheat Sheet

Category	Symbols	Example
Grouping	(), [], {}	`x = [1,2,3]`
Comparison	==, !=, >, <, >=, <=	`if x == 5:`

Category	Symbols	Example
Math	+, -, *, /, %, **, //	`total = 10 % 3`
Logic	and/or, &&,	
Assignment	=, +=, -=, *=	`x += 5`
Special	., ;, #, //, ->	`print(x.y)`

Final Takeaway

By thinking of programming symbols as **families**, you'll quickly see patterns:

- **Grouping** keeps code organized.
- **Math & Logic** handle calculations and decisions.
- **Bitwise & Assignment** give deeper control and shortcuts.
- **Comments & Special** keep code readable and language-specific.

This overview makes the rest of the book easier to navigate — you'll know exactly where a symbol fits in the bigger picture.

Language Symbol Families

Programming languages may seem like they speak different dialects, but most of them share the same alphabet of **symbols** — such as parentheses `()`, curly braces `{}`, brackets `[]`, and logical operators like `&&` or `||`.

What makes each language unique is **how it uses these symbols**.
Some are strict and symbolic (like C++), while others prefer clarity and readability (like Python).

Understanding these differences will help you confidently move between languages and avoid simple mistakes caused by confusing syntax.

🐍 The Python Family

What Is Python?

Python is a high-level, general-purpose programming language created by **Guido van Rossum** in 1991.
It was designed with one simple philosophy:

Code should be as easy to read as English.

Unlike older programming languages that rely heavily on punctuation and dense symbols, Python focuses on **clarity, simplicity, and readability.** Its clean design allows programmers to express powerful ideas with fewer lines of code.

Python is used almost everywhere — in **artificial intelligence**, **data science**, **web development**, **automation**, and even **education** — making it one of the most popular and beginner-friendly languages in the world.

Why Python?

Python has become the go-to choice for beginners and professionals because:

1. **It's easy to read and write.**
   The syntax feels natural, so you can focus on logic instead of memorizing symbols.
2. **It uses plain English words.**
   Instead of cryptic operators like `&&` or `!`, Python uses words like `and`, `or`, and `not`.
3. **It has consistent structure.**
   Indentation replaces curly braces `{}`, making code look neat and organized.
4. **It's extremely versatile.**
   You can build apps, analyze data, control robots, or train AI models — all with the same language.
5. **It's supported by a huge community.**
   Millions of tutorials, forums, and libraries exist to help you solve almost any problem.

In short, Python is **powerful enough for experts** but **gentle enough for beginners.**

Python's Symbol Style

Python belongs to a **"readable symbol family."**
It uses familiar symbols, but simplifies their use so you can focus on meaning instead of memorization.

1. Grouping and Structure

Python uses three main types of brackets:

- **Parentheses** `()` – For grouping expressions or calling functions.
- **Square brackets** `[]` – For accessing items in a list or array.
- **Curly braces** `{}` – For creating sets or dictionaries (key-value pairs).

Unlike many other languages, **Python does not use semicolons (`;`)** to end statements. Each new line naturally separates instructions.

Example:

```
fruits = ["apple", "banana", "cherry"]
print(fruits[1])  # Output: banana
```

2. Logical Operators

Instead of the symbols `&&`, `||`, or `!`, Python uses clear English words:

- `and` → both conditions must be true
- `or` → at least one condition must be true
- `not` → reverses a condition (True becomes False, and vice versa)

Example:

```
age = 25
if age >= 18 and age <= 30:
    print("Eligible to apply")
```

This line reads almost like plain English — and that's the beauty of Python.

3. Mathematical Operators

Python supports all standard arithmetic symbols (`+`, `-`, `*`, `/`, `%`) but also introduces two unique ones:

- `//` → **Floor Division**: divides numbers and rounds down to the nearest whole number.
- `**` → **Exponentiation**: raises one number to the power of another.

Example:

```
print(9 // 2)   # Output: 4
print(2 ** 4)   # Output: 16
```

These make Python more expressive and mathematically precise than many older languages.

4. Assignment Operators

Assignment operators allow you to update variables without rewriting long equations.

Symbol	Description	Example	Meaning
`=`	Assigns a value	`x = 5`	x becomes 5
`+=`	Adds and reassigns	`x += 3`	x = x + 3
`-=`	Subtracts and reassigns	`x -= 2`	x = x - 2
`*=`	Multiplies and reassigns	`x *= 4`	x = x × 4
`/=`	Divides and reassigns	`x /= 2`	x = x ÷ 2
`//=`	Floor divides and reassigns	`x //= 3`	x = floor(x ÷ 3)
`%=`	Modulus (remainder)	`x %= 2`	x = remainder of x ÷ 2
`**=`	Exponentiates and reassigns	`x **= 3`	x = x^3

11

These shortcuts make your code shorter and easier to maintain.

5. Comments and Documentation

Python uses the **hash symbol (#)** for single-line comments — anything after # on that line is ignored by the computer.
For longer explanations, Python allows **multi-line comments** called **docstrings**, created using triple quotes (""" """).

Example:

```
# This is a single-line comment

"""
This is a multi-line comment,
often used to describe how a function works.
"""
```

Comments are vital for writing clean, understandable code — both for yourself and others.

6. Special Symbols

Python uses a few unique symbols that make it stand out:

- **Colon** : — starts a new indented block (like a function or an if-statement).
- **At symbol** @ — used for **decorators**, which modify how a function behaves.
- **Type hints** — use a colon : to describe what kind of data a variable should hold.

Example:

```
def greet(name: str):
    # Simple greeting function
    return f"Hello, {name}"
```

Here, `name: str` tells the reader (and Python tools) that `name` should be a string.

7. Putting It All Together

Let's see how all of these ideas combine in one short, readable program:

```
def greet(name: str):
    # A simple greeting function
    return f"Hello, {name}"

age = 20

if age >= 18 and age <= 30:
    print(greet("Alice"))
```

What's happening here:

- The **colon** : marks where the code block starts.
- The **logical operators** (`and`, `<=`, `>=`) make the decision readable.
- The **function call** `greet("Alice")` uses parentheses to pass an argument.
- **No semicolons** are required — Python knows each line is a new instruction.

Summary: Why the Python Family Stands Out

The **Python Symbol Family** is built on the idea that **readability equals power.**
It removes unnecessary symbols, replaces them with words, and lets indentation and structure make the meaning clear.

Where other languages rely on symbols to control the code, Python relies on **logic, simplicity, and clean design.**
That's why it's not just a programming language — it's a teaching language, a research tool, and the foundation of modern AI and data science.

Python proves that **you don't have to write complicated code to do powerful things.**

🐾 *The Java Family*

What Is Java?

Java is a powerful, object-oriented programming language created by **James Gosling** at **Sun Microsystems** in 1995 (now owned by Oracle).
It was built on one core promise:

"Write once, run anywhere."

That means Java code can run on almost any device — computers, phones, or servers — without modification. This universal compatibility made Java one of the world's most widely used languages, powering millions of applications from **Android apps** to **enterprise banking systems**.

Java is considered a **strongly typed** and **structured language**, which means it enforces clear rules about how symbols, variables, and data types are used. This structure helps programmers write code that's consistent, reliable, and easy to maintain — even in massive projects.

Why Java?

Java remains a cornerstone of programming because it combines **discipline and flexibility**. Here's why it continues to be one of the most trusted languages for professionals worldwide:

13

1. **Cross-Platform Compatibility** – Code written once can run on any system that has the Java Virtual Machine (JVM).
2. **Strong Typing** – Every variable must have a defined type, which helps prevent errors before the program runs.
3. **Object-Oriented Design** – Java encourages thinking in terms of "objects" — reusable components that model real-world things.
4. **Stability and Security** – Java has built-in safeguards, making it a reliable choice for large organizations.
5. **Community and Libraries** – With decades of development, Java offers vast libraries, frameworks, and community support for nearly every kind of project.

In short, Java teaches programmers to **think carefully** about their code — a discipline that pays off in any other language they learn later.

Java's Symbol Style

Java belongs to the **"structured symbol family"**, where symbols define both **order and precision**.
Its syntax is similar to languages like C and C++, emphasizing the use of **braces, semicolons,** and **operators** to control how code executes.

1. Grouping and Structure

Java uses three primary grouping symbols — **parentheses ()**, **square brackets []**, and **curly braces {}** — **all of which are required** for organizing code properly.

- **Parentheses ()** define conditions or group expressions.
- **Square brackets []** represent arrays (ordered lists of values).
- **Curly braces {}** define *blocks of code* such as classes, methods, or loops.

In Java, **every block must be enclosed in braces**, and **every instruction must end with a semicolon (;)**.

Example:

```java
public class Main {
    public static void main(String[] args) {
        int age = 25;
        if (age >= 18 && age <= 30) {
            System.out.println("In range");
        }
    }
}
```

Here, {} contains the body of the class and method, while () groups the condition. The semicolon ends the System.out.println() statement.

2. Logical Operators

Java uses symbolic logic operators rather than words:

- `&&` → "and" (both conditions must be true)
- `||` → "or" (at least one condition must be true)
- `!` → "not" (reverses a condition)

Example:

```
if (age >= 18 && age <= 30) {
    System.out.println("Eligible");
}
```

Unlike Python, which uses English words like `and` and `or`, Java prefers symbolic operators. This is one reason Java code looks more mathematical and less conversational.

3. Mathematical Operators

Java includes the basic arithmetic symbols used across most programming languages: `+`, `-`, `*`, `/`, and `%`.

However, Java **does not support `**` for exponentiation** or `//` for floor division. To perform powers, Java uses the `Math.pow()` function, and division always returns a decimal result if needed.

Example:

```
System.out.println(10 % 3);        // Output: 1 (remainder)
System.out.println(Math.pow(2, 3)); // Output: 8.0
```

4. Assignment Operators

Like Python, Java supports shorthand assignment operators that combine calculation and assignment.

Symbol	Description	Example	Meaning
`=`	Assigns a value	`x = 5`	x becomes 5
`+=`	Adds and reassigns	`x += 3`	x = x + 3
`-=`	Subtracts and reassigns	`x -= 2`	x = x - 2
`*=`	Multiplies and reassigns	`x *= 4`	x = x × 4
`/=`	Divides and reassigns	`x /= 2`	x = x ÷ 2
`%=`	Finds remainder and reassigns	`x %= 3`	x = remainder of x ÷ 3

Each of these makes arithmetic cleaner and prevents repetitive code.

5. Comments and Documentation

Java supports **three types of comments**, each with a specific purpose:

- `//` – Single-line comment
- `/* */` – Multi-line comment
- `/** */` – Documentation comment (used to generate professional JavaDocs)

Example:

```
// Single-line comment
/* This is a multi-line comment */
/**
 * This is a documentation comment.
 * It can describe classes and methods.
 */
```

Documentation comments (`/** ... */`) are particularly important in Java because they integrate directly into automatic documentation tools — something unique to the language.

6. Special Symbols in Java

- **Semicolon** `;` – Ends every instruction. Forgetting it will cause a compilation error.
- **Angle brackets** `<>` – Define *generics*, which make code reusable for different data types.
- **At symbol** `@` – Marks *annotations*, which give metadata to classes or methods (for example, `@Override`).

Example:

```
import java.util.List;

public class Example {
    @Override
    public String toString() {
        return "Annotation example";
    }

    public static <T> void display(List<T> items) {
        for (T item : items) {
            System.out.println(item);
        }
    }
}
```

Here, `@Override` tells Java that this method redefines one from a parent class, while `<T>` shows that the `display()` method can accept any data type — a key feature of **generics**.

7. Putting It All Together

Below is a complete and simple example of Java's structure and symbol use:

```
public class Main {
    public static void main(String[] args) {
        int age = 25;

        if (age >= 18 && age <= 30) {
            System.out.println("In range");
        } else {
            System.out.println("Out of range");
        }
    }
}
```

Explanation:

- `public class Main` begins the program and must be wrapped in `{}`.
- `main()` is the entry point of the application — every Java program starts here.
- Logical operators `&&` and `<=` help control decision-making.
- Every line of executable code ends with a **semicolon**.

Summary: Why the Java Family Matters

The **Java Symbol Family** represents **discipline, structure, and reliability**.
Where Python focuses on readability, Java focuses on **precision and consistency**. Every symbol — from the curly brace `{}` to the semicolon `;` — defines strict boundaries that keep large-scale programs stable.

Java trains you to think like an engineer: to build systems that are clear, organized, and scalable.
It's the language of **logic and order**, and understanding its symbols gives you the foundation to master nearly every modern programming language that came after it.

◯ *The JavaScript Family*

What Is JavaScript?

JavaScript is the language of the web — the one that makes websites come alive.
Created in **1995 by Brendan Eich** at Netscape, JavaScript was originally designed to make web pages interactive. Over the years, it evolved far beyond simple animations and buttons.

Today, JavaScript powers everything from **modern websites and mobile apps** to **server-side programs** and even **AI applications**. It's one of the most widely used languages in the world because it runs directly in your browser — no installation or setup required.

Where Java is strict and structured, **JavaScript is flexible and forgiving**. It allows developers to experiment freely, making it perfect for beginners and creative programmers alike.

Why JavaScript?

JavaScript is everywhere — if you've used the internet today, you've already used JavaScript. Here's why it's so essential:

1. **It runs in every browser.** No special tools needed.
2. **It's extremely flexible.** You can build interactive websites, games, servers, or even robots.
3. **It's easy to learn but powerful to master.** You can start small and grow your skills endlessly.
4. **It's constantly evolving.** Modern JavaScript (ES6 and beyond) adds new, elegant features like arrow functions and template strings.
5. **It connects the world.** JavaScript bridges design (frontend) and logic (backend), making it the universal language of the web.

JavaScript's Symbol Style

JavaScript belongs to the **"dynamic symbol family."**
It uses many of the same symbols as C and Java but allows greater flexibility — such as skipping semicolons in many cases (though adding them is still considered good practice).

1. Grouping and Structure

JavaScript uses:

- **Parentheses** `()` for grouping expressions and calling functions.
- **Square brackets** `[]` for arrays (lists of items).
- **Curly braces** `{}` for objects and code blocks (functions, loops, etc.).

Example:

```
function greet(name) {
    console.log(`Hello, ${name}`);
}
```

Here, `()` wraps the input, `{}` contains the code block, and the **backticks** (`` ` ``) are part of JavaScript's *template strings*.

2. Logical Operators

JavaScript uses standard logic symbols, similar to Java and C:

- `&&` → and

18

- | | → or
- ! → not

Example:

```
let age = 28;
if (age >= 18 && age <= 30) {
    greet("Alice");
}
```

Just like in Java, both `&&` and `||` evaluate conditions, but JavaScript lets you mix logic and data types more freely — something called **type coercion**.

3. Mathematical Operators

JavaScript supports all common arithmetic operators:
`+`, `-`, `*`, `/`, and `%`.
In modern versions (ES6+), it also supports:

- `**` → **Exponentiation** (raises one number to the power of another).

For floor division (rounding down), JavaScript uses the built-in method:
`Math.floor()`

Example:

```
console.log(9 / 2);          // Output: 4.5
console.log(Math.floor(9 / 2)); // Output: 4
console.log(2 ** 3);         // Output: 8
```

4. Assignment Operators

JavaScript includes assignment shortcuts just like Python and Java.

Symbol	Description	Example	Meaning
=	Assigns a value	x = 5	x becomes 5
+=	Adds and reassigns	x += 3	x = x + 3
-=	Subtracts and reassigns	x -= 2	x = x - 2
*=	Multiplies and reassigns	x *= 4	x = x × 4
/=	Divides and reassigns	x /= 2	x = x ÷ 2
%=	Modulus and reassigns	x %= 3	x = remainder of x ÷ 3

5. Comments

JavaScript uses two types of comments:

- `//` for single-line comments
- `/* */` for multi-line comments

Example:

```
// This is a single-line comment
/* This is a multi-line comment */
```

Comments are ignored by the browser but essential for explaining your code clearly.

6. Special Symbols and Modern Features

Modern JavaScript introduces new, expressive symbols:

- **Backticks (`` ` ``)** – used for **template strings**, allowing variables inside text.
- **Optional chaining (?.)** – safely access nested data without errors.
- **Arrow functions (=>)** – shorter, cleaner syntax for writing functions.

Example:

```
const greet = (name) => {
    console.log(`Hello, ${name}`);
};

let user = { profile: { name: "Alice" } };
console.log(user?.profile?.name);  // Optional chaining prevents errors
```

Summary: Why the JavaScript Family Matters

The **JavaScript Symbol Family** represents **freedom, flexibility, and creativity**.
It blends structure with simplicity — allowing you to write powerful programs without too much rigidity.

Where Python feels like plain English and Java feels like formal grammar, JavaScript feels like **conversation** — open, fast, and expressive.

It's the language that built the modern web — and understanding its symbols means understanding how the digital world communicates.

🖥️ The C & C++ Family

What Are C and C++?

C (created by **Dennis Ritchie** in the early 1970s) is the ancestor of most modern programming languages. It was built for **speed, control, and efficiency** — giving programmers direct access to computer memory.

C++ (created by **Bjarne Stroustrup** in 1985) extended C by adding **object-oriented programming**, allowing developers to structure code in reusable "objects."

Together, they form the **foundation of modern software engineering.**
Operating systems, browsers, game engines, and even AI systems use C or C++ underneath.

Why C and C++?

C and C++ remain vital because they teach you how computers truly think.
They don't hide complexity behind shortcuts — they *show you the machine*.

1. **Low-Level Control** – You can manage memory directly using pointers.
2. **Speed and Performance** – C/C++ programs run faster than most modern languages.
3. **Portability** – C and C++ code can run on nearly any device or operating system.
4. **Legacy Power** – Many critical systems (like operating systems or databases) are still written in C or C++.
5. **Foundation for Learning** – Mastering C/C++ helps you understand every other language more deeply.

C & C++ Symbol Style

The C family belongs to the **"low-level control symbol family."**
Every symbol carries weight — a missing semicolon or brace can crash an entire program.

1. Grouping and Structure

C and C++ rely heavily on structure:

- **Parentheses** () for grouping and function calls.
- **Square brackets** [] for arrays.
- **Curly braces** {} for defining code blocks.
- **Semicolon** ; ends every statement.

Example:

```
#include <iostream>
using namespace std;

int main() {
    int x = 5;
    int* p = &x; // pointer
    if (x > 0 && *p == 5) {
        cout << "Positive five" << endl;
    }
    return 0;
}
```

2. Logical Operators

C and C++ use classic logic symbols:

- `&&` → and
- `||` → or
- `!` → not

They behave exactly as in Java — precise, fast, and strict about data types.

3. Mathematical Operators

These languages include the familiar operators:
`+, -, *, /,` and `%`.
For exponentiation, you must use the `pow()` function from the `<math.h>` library.

Example:

```
#include <cmath>
cout << pow(2, 3); // Output: 8
```

4. Assignment Operators

Assignment operators behave the same way as in Java:
`=, +=, -=, *=, /=,` and `%=`.

Each one updates and reassigns a value directly to a variable.

5. Comments

C and C++ use:

- `//` for single-line comments
- `/* */` for multi-line comments

6. Special Symbols

These languages introduce several powerful symbols that control how memory and data are handled:

- `*` – Pointer (holds a memory address).
- `&` – Address-of operator (returns a variable's memory location).
- `->` – Accesses members of a structure or class through a pointer.
- `::` – Scope resolution operator (accesses global or class-level identifiers).

Example:

```
int x = 10;
```

```
int* p = &x;    // Pointer to x
cout << *p;     // Prints the value of x
```

These are what give C and C++ their "machine-level" power — and also why they require extra care.

Summary: Why the C & C++ Family Matters

The **C and C++ Symbol Family** represents **power, precision, and control**.
Every symbol has meaning — every semicolon and pointer teaches you how computers truly work.

While Python and JavaScript simplify programming, C and C++ expose its raw structure. They form the backbone of the digital world — the silent engines behind operating systems, databases, compilers, and embedded devices.

Learning their symbols gives you not just programming skill, but **engineering insight** — the kind that separates coders from true computer scientists.

⚡ The C# Family

What Is C# (pronounced "C-Sharp")?

C# was developed by **Microsoft** in 2000 as part of its .NET Framework.
It was designed to combine the **discipline of Java** with the **power of C++**, while still being modern, safe, and easy to read.

C# quickly became the language of choice for **Windows applications**, **enterprise software**, **Unity game development**, and **cloud services** on Azure.
It's an **object-oriented, strongly-typed** language — meaning every symbol and value must have a clearly defined type, which keeps your programs stable and predictable.

Why C# Matters

C# balances **power and safety**.
It lets developers build complex systems without the low-level memory risks of C or C++.
Key reasons why C# remains a top choice:

1. **Clean, structured syntax** – looks like Java but feels smoother.
2. **Null safety features** – operators like `?.` and `??` help prevent crashes from missing data.
3. **Rich tooling** – Visual Studio and .NET make debugging and deployment simple.
4. **Cross-platform support** – via .NET Core, C# runs on Windows, macOS, and Linux.
5. **Modern features** – lambda expressions, async programming, and pattern matching keep the language fresh and powerful.

C# is often described as **the best of both worlds** — structured like Java but modern like JavaScript.

C# Symbol Style

C# belongs to the **"modern typed symbol family."**
It shares Java's core structure (`()`, `{}`, `;`) but adds its own advanced operators to handle null values, text, and functions more elegantly.

1. Logical and Null Operators

C# includes the classic logical operators from Java:

- `&&` → and
- `||` → or
- `!` → not

But it also introduces two powerful *null-safety* operators:

- `?.` → **Null-Conditional Operator** (safely access a member only if the object is not null).
- `??` → **Null-Coalescing Operator** (provides a default value if the expression is null).

Example:

```
int? age = null;
Console.WriteLine(age ?? 18); // Prints 18 if age is null
```

Here, `int?` declares a nullable integer, and `??` ensures the program never crashes when `age` is missing.

2. Math and Assignment Operators

C# uses the same arithmetic and assignment operators as Java:
`+, -, *, /, %, =, +=, -=, *=, /=, %=.`
Each operator performs exactly as you'd expect — but C#'s compiler strictly checks types, reducing logic errors.

3. Special Symbols

C# adds a few symbols that make it feel modern and expressive:

- `=>` – **Lambda Expressions**, used for short functions and inline logic.
- `@` – **Verbatim Strings**, which allow multi-line text without escaping characters.

Example:

```
var greet = (name) => $"Hello, {name}";
Console.WriteLine(greet("Alice"));

string path = @"C:\Users\Documents\file.txt"; // @ ignores backslashes
```

Summary: Why the C# Family Stands Out

The **C# Symbol Family** represents **clarity, safety, and modern design**.
It builds on Java's discipline but adds null-safety and expressive features that make programming more intuitive.
If you enjoy structure but crave flexibility, C# is the perfect balance — a language that grows with you as your projects expand from desktop to cloud to game development.

◯ The HTML & CSS Family

What Is HTML and CSS?

HTML (HyperText Markup Language) and **CSS (Cascading Style Sheets)** are the two core technologies of the web.
Think of HTML as the **skeleton** that gives a web page structure and meaning, and CSS as the **skin and style** that make it look beautiful.

HTML defines *what* appears on a page — headings, paragraphs, images, and links.
CSS defines *how* it appears — colors, fonts, layouts, and animations.

Together, they form the visual language of the internet.

HTML — Structure

HTML uses **tags**, **attributes**, and **comments** as its primary symbols.

1. Tags

Tags mark the start and end of elements. They're wrapped in angle brackets < >:

```
<h1>Hello World</h1>
```

Here, `<h1>` starts a heading and `</h1>` ends it.

2. Attributes

Attributes add details to tags and use the = symbol to assign values:

```
<a href="https://example.com">Visit Website</a>
```

In this example, `href` defines the link's destination.

3. Comments

Comments begin with `<!--` and end with `-->`.
They help developers explain their code without affecting the page.

Example of a complete HTML page:

```
<!DOCTYPE html>
<html>
<head>
  <title>My Page</title>
</head>
<body>
  <!-- This is a comment -->
  <h1>Hello World</h1>
  <p>Welcome to my first website!</p>
</body>
</html>
```

CSS — Styling

CSS controls appearance using **selectors**, **properties**, and **values**.

1. Selectors

Selectors target specific HTML elements.

- . (dot) for class names → `.intro { }`
- # (hash) for IDs → `#main { }`
- Element names like `h1`, `p`, or `div`.

2. Braces, Colons, and Semicolons

- Curly braces `{ }` contain style rules.
- Colons `:` separate a property from its value.
- Semicolons `;` end each rule.

Example:

```
h1 {
  color: blue;
  font-size: 24px;
}
```

3. Special Directives

CSS includes special symbols like `@media` and `@keyframes` to create responsive and animated designs.

Example:

```
@media (max-width: 600px) {
  h1 { font-size: 18px; }
}

@keyframes fadeIn {
  from { opacity: 0; }
  to { opacity: 1; }
}
```

Summary: Why the HTML & CSS Family Matters

The **HTML and CSS Symbol Family** forms the foundation of every web page you've ever seen.
HTML gives structure; CSS adds beauty and motion.

Together, they teach you how symbols can create not just logic, but **visual communication** — turning raw text and code into interactive art.

Once you understand their symbols, you understand how the internet itself is built.

The SQL Family

What Is SQL?

SQL (Structured Query Language) is the language of databases.
Instead of giving step-by-step commands, you *ask* the database for what you want — and it retrieves it for you.
That's why SQL is known as a **declarative language** — you declare *what* you want, not *how* to get it.

Created in the 1970s, SQL is used across nearly every industry for storing, organizing, and analyzing data. Whether you're checking bank transactions, searching an online store, or managing a website's user accounts, SQL is working quietly in the background.

27

Why SQL?

SQL remains vital because it gives humans a clear, structured way to communicate with machines about data.

1. **Universal Database Language** – Works with MySQL, PostgreSQL, Oracle, and most database systems.
2. **Readable Syntax** – Commands like SELECT, FROM, and WHERE read like plain English.
3. **Powerful Filtering** – Retrieve exactly the data you need with conditions and operators.
4. **Data Control** – Insert, update, delete, and manage millions of records efficiently.
5. **Essential for AI and Analytics** – Every data-driven system starts with organized data, and that's SQL's specialty.

SQL's Symbol Style

SQL belongs to the **"data query symbol family."**
It uses mathematical and logical symbols to define relationships between data values and textual keywords to describe actions.

1. Math and Logic Operators

SQL supports familiar logical and comparison symbols:

- = → equals
- <> → not equal
- > and < → greater than / less than
- >= and <= → greater than or equal / less than or equal
- AND, OR, NOT → logical connectors

Example:

```
SELECT name, age
FROM users
WHERE age >= 18 AND age <= 30;
```

This command selects the names and ages of users whose ages are between 18 and 30.

2. Special Symbols

SQL uses a few symbols that make its queries powerful and compact:

- * → selects all columns (e.g., SELECT * FROM users;)
- % → **wildcard symbol**, used with LIKE to find similar patterns (e.g., WHERE name LIKE 'A%')
- ; → ends a statement (especially when running multiple commands).

Example:

```
SELECT *
FROM products
WHERE name LIKE 'C%';
```

This retrieves all products whose names begin with the letter "C."

Summary: Why the SQL Family Matters

The **SQL Symbol Family** teaches **precision and logic** through simplicity.
It proves that programming doesn't always require loops or functions — sometimes, it's about asking the right question in the right format.

SQL gives programmers, analysts, and AI systems access to the world's most valuable resource: **data**.

⬤ The Web Scripting & Markup Family (Extra)

Modern technology doesn't end with code that runs — it also includes the languages that **display, describe, and structure information**.
This family includes lightweight formats and markup languages used to **format text**, **transfer data**, and **configure systems**.

Markdown

Markdown is a simple text-formatting language designed for readability.
It's used in documentation, blogs, and GitHub files because it converts plain text into HTML-like structure without clutter.

Symbol	Purpose	Example
#	Headings	# Title, ## Subtitle
** **	Bold text	**important**
[]()	Hyperlinks	[Visit Site](https://example.com)

Example:

```
# Welcome
**This text is bold**, and here's a [link](https://example.com).
```

JSON (JavaScript Object Notation)

JSON is a lightweight data format used to transfer structured information between systems. It's built from **objects** {} and **arrays** [], with key/value pairs separated by colons : and commas , .

Example:

```
{
  "name": "Alice",
  "age": 25,
  "skills": ["Python", "JavaScript", "SQL"]
}
```

JSON is readable, universal, and used everywhere — from APIs to AI data exchange.

YAML (YAML Ain't Markup Language)

YAML is a configuration format known for its clean, indentation-based structure.
It's commonly used in **DevOps**, **AI model configuration**, and **cloud systems**.

Example:

```
name: Alice
age: 25
skills:
  - Python
  - JavaScript
  - SQL
```

Unlike JSON, YAML uses indentation (spaces) instead of braces, making it easier to read but sensitive to formatting.

Summary: Why the Web Scripting & Markup Family Matters

The **Web Scripting and Markup Symbol Family** connects humans and machines through **structured communication**.
Markdown simplifies text formatting, JSON organizes data, and YAML structures configuration — all using simple, predictable symbols.

They form the invisible backbone of modern web services, making digital systems easy to share, sync, and scale.

🥔 Quick Comparison Table

Feature / Symbol	Python	Java	JavaScript	C / C++	HTML / CSS	SQL
AND	and	&&	&&	&&	—	AND
OR	or	`		`	`	
NOT	not	!	!	!	—	NOT
Exponentiation	**	Math.pow()	**	pow()	—	—
Floor Division	//	—	Math.floor()	—	—	—
Comments	#	//, /* */	//, /* */	//, /* */	<!-- -->, /* */	--, /* */
Blocks	: + indent	{}	{}	{}	{} (in CSS)	—
Statement End	Newline	;	Optional ;	;	; (in CSS)	;

💡 Why Add This Section

Including this **Language Comparison Section** helps your readers see programming languages not as isolated systems, but as **different dialects of one universal coding language**.

☑ **It gives context:** Readers can visually compare symbols across multiple languages.
☑ **It builds confidence:** They can transition from one language to another more easily.
☑ **It supports understanding:** Seeing similarities reinforces how symbols carry shared logic.
☑ **It fits perfectly:** This section flows naturally after your *Comprehensive Symbol Reference* and before your *Final Exercises*.

In short, this section transforms your book from a reference manual into a **bridge between programming languages** — helping every learner see how different code families speak the same symbolic language.

🧩 The PHP Family

What Is PHP?

PHP (Hypertext Preprocessor) is a server-side scripting language created by **Rasmus Lerdorf** in 1994.
Originally designed to track website visits, PHP evolved into one of the most widely used languages for building **dynamic web applications**.

31

Unlike HTML, which only displays content, **PHP can think** — it processes data, interacts with databases, manages forms, and generates web pages dynamically before they're sent to your browser.

Whenever you fill out a form, sign in to a website, or make an online purchase, there's a good chance PHP is working quietly in the background.

Why PHP?

Despite being nearly three decades old, PHP remains one of the **pillars of web development**, powering over **70% of all websites**, including WordPress, Facebook (in its early days), and Wikipedia.

Here's why PHP is still essential:

1. **Runs on the server.** PHP executes on web servers (like Apache or Nginx), making it ideal for secure processing.
2. **Easily connects to databases.** MySQL, MariaDB, and PostgreSQL integrate seamlessly.
3. **Blends with HTML.** PHP and HTML can live together in the same file.
4. **Flexible and open source.** It runs on almost any platform — Windows, Linux, or macOS.
5. **Fast and mature.** Decades of optimization make PHP reliable, stable, and efficient for large-scale systems.

In short, PHP is the **engine** behind many websites — practical, powerful, and deeply embedded in the web's foundation.

PHP's Symbol Style

PHP belongs to the **"server-side symbol family."**
Its symbols are a blend of **C-style syntax** (like ; and { }) and **unique identifiers** like $ for variables and -> for object references.

PHP files usually start with the <?php tag and end with ?>, signaling that the enclosed code should be interpreted by the PHP engine.

1. Variables and Data Symbols

In PHP, **all variables begin with the dollar sign $**, followed by the variable name.

Example:

```
<?php
$name = "Alice";
$age = 25;
?>
```

Here:

- $name and $age are variables.
- The = symbol assigns values (just like in other languages).
- Every statement ends with a **semicolon** ;.

2. Grouping and Structure

PHP uses the same structural symbols as Java and C:

- **Parentheses ()** for conditions and function calls.
- **Curly braces {}** for defining code blocks.
- **Semicolons** ; to end each statement.

Example:

```php
<?php
if ($age >= 18 && $age <= 30) {
    echo "Welcome, $name!";
}
?>
```

Here:

- The parentheses () group the condition.
- {} define the block of code to run if the condition is true.
- echo outputs text to the browser.

3. Logical and Comparison Operators

PHP uses logic symbols similar to Java and JavaScript:

- && → and
- || → or
- ! → not

And standard comparison operators:

- == → equal to
- === → identical (same value and type)
- != or <> → not equal
- > < >= <= → numeric comparisons

Example:

```
if ($name == "Alice" && $age >= 18) {
    echo "Access granted.";
}
```

4. Arrays and Objects

PHP uses several key symbols to define complex data:

Symbol	Purpose	Example
[]	Create arrays	$fruits = ["apple", "banana", "cherry"];
=>	Map keys to values in associative arrays	$person = ["name" => "Alice", "age" => 25];
->	Access object properties or methods	$user->login();

Example:

```
<?php
$person = ["name" => "Alice", "age" => 25];
echo $person["name"]; // Prints: Alice
?>
```

5. Math and Assignment Operators

PHP supports familiar math symbols:
+, -, *, /, %, and ** (for exponentiation).

And assignment shortcuts:
=, +=, -=, *=, /=, and %=.

Example:

```
<?php
$x = 10;
$x += 5;    // Same as $x = $x + 5
echo $x;    // Prints: 15
?>
```

6. Comments

PHP supports both single-line and multi-line comments:

- // → single line

- `/* ... */` → multi-line

Example:

```php
<?php
// This is a single-line comment
/* This is
   a multi-line comment */
?>
```

7. Strings and Output Symbols

PHP uses several ways to display or format text:

- `echo` – Prints text or variables to the screen.
- `.` – Concatenates (joins) strings.
- **"** and **'** – Used for strings (double quotes interpret variables; single quotes don't).

Example:

```php
<?php
$name = "Alice";
echo "Hello, " . $name . "!";  // Prints: Hello, Alice!
?>
```

8. Putting It All Together

Below is a complete PHP example showing how symbols combine to form logic:

```php
<?php
$name = "Alice";
$age = 28;

function greet($person) {
    return "Hello, " . $person . "!";
}

if ($age >= 18 && $age <= 30) {
    echo greet($name);
} else {
    echo "Access denied.";
}
?>
```

Explanation:

- `function` defines reusable code.
- `()` holds parameters.
- `{}` defines the code block.

- `$` defines variables.
- `&&` connects conditions.
- `;` ends each statement.

Summary: Why the PHP Family Matters

The **PHP Symbol Family** represents **interaction, flexibility, and real-time data handling**. It bridges logic with the web — enabling developers to create dynamic pages that respond instantly to user input.

Where HTML structures content and JavaScript adds interactivity, PHP controls the *brains* behind the website — the part that connects your browser to a server, database, or API.

Understanding PHP's symbols gives you the power to build websites that don't just display information — they *think*, *respond*, and *adapt*.

🪨 *The Swift Family*

What Is Swift?

Swift is Apple's modern programming language, introduced in **2014** by **Apple Inc.** as a replacement for **Objective-C**.
It was designed to be **fast, safe, and easy to read**, making it one of the most beginner-friendly languages for app development.

Swift powers nearly everything in the **Apple ecosystem** — including **iOS**, **macOS**, **watchOS**, and **tvOS**.
Whether you're building an iPhone app, a Mac tool, or even an AI-driven project using Apple's frameworks, Swift is the language that makes it all possible.

Why Swift?

Swift was created to combine **the power of C and Objective-C** with **the simplicity of Python**.
It's designed to be expressive, readable, and secure — while offering top-tier performance.

Here's why developers love Swift:

1. **Readable and elegant syntax** – It feels close to natural language.
2. **Safety first** – Swift prevents common coding errors like null pointer crashes.
3. **Speed and performance** – Swift is built for Apple's hardware, making it incredibly fast.
4. **Cross-platform potential** – With Swift open-sourced, it now runs on Linux and Windows too.
5. **Future-ready** – It integrates seamlessly with AI, AR, and advanced iOS frameworks.

Swift is not just a coding language — it's Apple's vision of the future of app development.

Swift's Symbol Style

Swift belongs to the **"safe and expressive symbol family."**
It borrows structure from C-style languages (`{}`, `()`, `;`) but simplifies and modernizes many concepts.
In Swift, clarity always comes before complexity — every symbol is chosen to make code easier to read and safer to execute.

1. Grouping and Structure

Swift uses the same familiar structure symbols as other languages:

- **Parentheses** `()` → For function calls and conditions.
- **Curly braces** `{}` → For code blocks.
- **Semicolons** `;` → Usually optional (Swift automatically ends lines).

Example:

```
if age >= 18 && age <= 30 {
    print("In range")
}
```

Here, the parentheses enclose the condition, and the braces define the block of code to execute.

2. Variables and Constants

Swift uses two key symbols for variable declaration:

- `var` → Creates a **variable** (a value that can change).
- `let` → Creates a **constant** (a value that cannot change).

Example:

```
var name = "Alice"
let birthYear = 1995
```

This approach prevents accidental changes to important data — a cornerstone of Swift's safety model.

3. Logical and Comparison Operators

Swift supports the standard logical and comparison operators:

- `&&` → and
- `||` → or
- `!` → not
- `==` → equal
- `!=` → not equal
- `> < >= <=` → numeric comparisons

Example:

```
if age >= 18 && age <= 30 {
    print("Eligible")
}
```

4. Math and Assignment Operators

Swift includes the usual arithmetic symbols:
`+, -, *, /, %` — plus **exponentiation** using built-in functions like `pow()`.

Assignment shortcuts are identical to those in C and Java:
`=, +=, -=, *=, /=`, and `%=`.

Example:

```
var x = 10
x += 5
print(x) // Output: 15
```

5. Optionals and Null Safety

Swift introduces a unique and powerful symbol system called **optionals**, which help prevent errors caused by missing values.

Symbol	Meaning	Example
?	Optional — value may exist or be nil	`var age: Int? = nil`
!	Force unwrap — assumes value exists	`print(age!)`
??	Default value if nil	`print(age ?? 18)`

Example:

```
var age: Int? = nil
print(age ?? 18)  // Prints 18 if age is nil
```

This feature makes Swift one of the safest programming languages in the world.

6. Strings and Interpolation

Swift makes text handling easy with **string interpolation**, allowing you to insert variables directly inside text using \().

Example:

```
var name = "Alice"
print("Hello, \(name)!")
```

Output:

```
Hello, Alice!
```

7. Functions and Closures

Swift introduces **arrow syntax (->)** for defining what a function returns.
It also supports **closures** (similar to JavaScript's arrow functions) for compact inline logic.

Example:

```
func greet(person: String) -> String {
    return "Hello, \(person)!"
}

let greetUser = { (name: String) -> String in
    return "Hi, \(name)!"
}
```

Here, the `->` `String` means the function returns a string value.

8. Special Symbols

Swift adds several symbols that give it power and expressiveness:

- `:` → Declares data types (`var age: Int = 25`)
- `->` → Defines return types for functions
- `?` / `!` → Handle optionals and null safety
- `@` → Used for attributes like `@IBOutlet` or `@objc`
- `.` → Accesses properties or methods (`person.name`)

Example:

```
class Person {
    var name: String
    init(name: String) {
        self.name = name
    }
}
```

```
let user = Person(name: "Alice")
print(user.name)
```

9. Putting It All Together

Here's a complete Swift example that shows most of these symbols in action:

```
func greet(person: String?, age: Int?) {
    let userName = person ?? "Guest"
    let userAge = age ?? 18

    if userAge >= 18 && userAge <= 30 {
        print("Welcome, \(userName)! You're \(userAge) years old.")
    } else {
        print("Access restricted for age \(userAge).")
    }
}

greet(person: "Alice", age: 25)
```

Explanation:

- ?? provides default values for missing data.
- () groups function parameters.
- {} encloses code blocks.
- \() interpolates variables in text.
- ; is optional.

Summary: Why the Swift Family Matters

The **Swift Symbol Family** represents **safety, clarity, and evolution**.
It combines the readability of Python, the structure of Java, and the precision of C — all wrapped in Apple's sleek design philosophy.

Every symbol in Swift serves a purpose: to make coding faster, safer, and more enjoyable. By learning Swift's syntax and symbols, you gain the ability to create world-class apps that power iPhones, Macs, and beyond — all while writing code that reads like clear, human language.

Swift is the language of the future — elegant, intelligent, and built to last.

"I wrote this book while learning to code because I couldn't find a resource that explained the symbols themselves. After struggling through confusing tutorials, I realized what was missing: a clear guide to what all those brackets, operators, and special characters actually mean. This is the book I wish existed when I started."

Why We Need Symbols in Coding

Programming is a language. Just as human languages use punctuation and grammar to create meaning, **programming languages use symbols** to tell the computer what to do. These symbols are the smallest, fastest way to express complex ideas.

1. They Are the Building Blocks of Logic

- Symbols such as =, ==, >, <, &&, and || let us **compare values and make decisions**.
- Without them, you couldn't write conditions like:

```
if score >= 100:
    print("You win!")
```

Here, >= instantly tells the computer: *greater than or equal to*.

2. They Control How Data Is Stored and Changed

- Assignment symbols like = and compound forms such as += or *= let you **store and update information** quickly.
- Example:

```
let points = 0;
points += 10; // adds 10 points
```

3. They Power Calculations and Transform Data

- Math symbols + - * / % ** let you calculate everything from shopping totals to scientific formulas.
- Example:

```
area = width * height
discount_price = price - (price * 0.2)
```

4. They Organize Code into Meaningful Structures

- Grouping symbols — parentheses (), brackets [], braces {} — organize your code.
- They decide **where functions begin and end, how arrays look, and which code belongs inside loops or conditions**.

```
for (int i = 0; i < 5; i++) {
    System.out.println(i);
}
```

5. They Make Code Shorter, Clearer, and Faster

- Instead of long words, symbols **compress complex actions into one character**.
- For example, `&&` is faster and cleaner than writing "if both conditions are true."

6. They Enable Communication Across All Languages

- Symbols are a **universal part of coding** — once you know them in Python, you'll recognize them in JavaScript, Java, C#, and more.
- Learning them first makes switching languages easier.

☑ Takeaway

Symbols are the punctuation, math signs, and traffic signals of code.
They let programmers build logic, perform calculations, control flow, and organize ideas — all while keeping code short and powerful.
Understanding symbols early will save you hours of debugging and help you write code that's both **correct and elegant**.

Where Do Programmers Write Code?

Programmers don't write code inside the computer's operating system directly — they use **special environments and tools** designed to make coding easier, safer, and more organized.

1. Code Editors (Lightweight Writing Tools)

These are like a super-powered Notepad made for coding. They highlight symbols, check for errors, and make text easier to read.

- **Examples:**
    - Visual Studio Code (VS Code)
    - Sublime Text
    - Atom
- **Why use:** Great for quick projects and beginners.

```
hello_world.py  ← file written in VS Code
```

2. Integrated Development Environments (IDEs)

Full-featured tools with **editing, running, debugging, and project management** built in. Ideal for larger projects.

- **Examples:**
    - PyCharm (Python)
    - IntelliJ IDEA (Java)

- Eclipse (Java/C++)
- Visual Studio (C#, C++)
- **Why use:** They help you catch mistakes early, run code, and manage big programs easily.

3. Online Coding Platforms

Websites where you can code directly in your browser — no installation required.

- **Examples:**
  - Replit (many languages)
  - CodePen (HTML/CSS/JS)
  - JSFiddle (JavaScript)
  - Google Colab (Python + data science)
- **Why use:** Perfect for practice, quick demos, and sharing code with others.

4. Command Line / Terminal

Advanced users sometimes write and run code directly in the **terminal** or a **shell** (like Bash, Zsh, or PowerShell).

- **Examples:**
  - Writing Python scripts and running them with `python myfile.py`
  - Compiling Java with `javac MyClass.java`
- **Why use:** Fast and powerful for experienced programmers, automation, and servers.

5. Cloud Development Environments

Full coding environments hosted online, often used by teams.

- **Examples:**
  - GitHub Codespaces
  - AWS Cloud9
  - Gitpod
- **Why use:** Code anywhere with just a browser, no setup on your computer.

☑ Takeaway

Programmers **don't just use one tool**.

- Beginners often start with **online platforms** or **simple editors** like VS Code.
- Professionals use **IDEs** for big projects and sometimes the **terminal** for quick commands.
- Teams may use **cloud environments** so everyone works on the same setup.

Think of coding spaces like workshops: from a small desk (simple editor) to a full workshop with all the tools (IDE), to a shared online workspace (cloud).

A Note from the Author

This book was born out of countless moments of curiosity, confusion, and determination. When I first began learning to code, I remember staring at the screen and wondering why no one explained what all those little symbols meant. Everyone seemed to assume you already knew. But I didn't—and that missing piece made learning feel harder than it needed to be.

I'm not a computer science professor or a lifelong programmer. I'm simply someone who fell in love with the logic behind code and wanted to make it easier for others to understand. Along the way, I realized that many beginners struggle not because they lack ability, but because no one ever takes the time to explain *why* symbols behave the way they do.

Every page of this book was written with that learner in mind. Each example was tested, refined, and simplified to ensure that you can follow along confidently—even if you're seeing these concepts for the first time. My goal was to make programming less intimidating and more human—something anyone can learn, step by step.

That said, programming is always changing. Languages evolve, tools improve, and even experts continue to learn. So, I encourage you to:

- **Experiment** with every code sample. Try changing numbers, words, or symbols to see what happens.
- **Explore** your language's official documentation whenever you're unsure.
- **Connect** with online coding communities and share your journey with others.
- **Use** this book as a launchpad—a place to build your foundation, not your finish line.

If this book helps even one reader overcome the same confusion I once felt, then every hour spent writing it was worth it. My hope is that it gives you the clarity and confidence to keep moving forward—because once you understand how code *speaks,* the rest of programming becomes much easier to learn.

Thank you for letting me be part of your learning journey. I'm walking this path right alongside you—and cheering you on with every page you turn.

Happy coding,
Susie Hala

First Edition — October 2025
Last Updated — October 2025
Contact: halaprosper1@gmail.com

"I wrote this book while learning to code because I couldn't find a resource that explained the symbols themselves. After struggling through confusing tutorials, I realized what was missing: a clear guide to what all those brackets, operators, and special characters actually mean. This is the book I wish existed when I started."

Why This Book Is Essential for Every Aspiring Programmer

The Missing Foundation That No One Teaches

If you've ever tried learning programming through online tutorials, YouTube videos, or bootcamps, you've probably encountered the same frustrating pattern.

The instructor smiles and says:

"Now, let's use an **if-statement** to check the user's age."

Then types something like this:

```
if (user_age >= 18) && (has_license == True):
    print("You can drive!")
```

You copy the code exactly. But it doesn't work.

You stare at the screen, puzzled—
What's the difference between = and ==?
Why are there two & symbols?
What do those parentheses actually do?

Meanwhile, the instructor has already moved on, assuming you knew all that.

That's the hidden **gap** that causes most beginners to give up.

Traditional programming lessons often focus on *what code does*—not *how* or *why* it works at the symbol level. They rush into topics like loops, functions, or variables, assuming you already understand that != means "not equal" or that [] and {} serve entirely different purposes.

But here's the truth: **no one ever teaches you where to learn the meaning of the symbols themselves**—the smallest, most powerful pieces of programming.

What Makes This Book Different

This book does what most tutorials skip—it teaches you the **language of symbols** that gives code its meaning.

Before diving into complex topics, we start at the real beginning: **understanding the role of every bracket, operator, and special character** that holds programs together.

Inside, you'll find clear answers to the questions that usually go unasked:

- Why do some things use () while others use [] or {}?
- What's the difference between =, ==, and ===?
- Why does my code break when I use & instead of &&?
- What does -> do differently from . or ::?
- When do I need semicolons—and when can I skip them?

These are not minor details. They are the *core grammar* of programming—the foundation on which every concept is built.

Once you understand the meaning, structure, and behavior of symbols, the rest of programming—loops, logic, data types, and functions—suddenly makes perfect sense.

That's the power of this book: it helps you see **how code really speaks**.

A Message Before You Begin

Every great programmer starts by understanding the smallest details.
Brackets, operators, and punctuation may look simple, but they are the alphabet of logic—the symbols that turn imagination into instruction.

If you take the time to master what each symbol means, you'll gain not only technical skill but also **clarity, confidence, and creativity**.

So, take a deep breath, open your mind, and get ready to see programming in a completely new light.
The next page marks the beginning of your journey—from *confusion to comprehension*.

What Makes This Book Different

This book teaches what everyone else skips: the actual symbols of programming.

Instead of jumping straight to loops and functions, we start at the true beginning—understanding what every bracket, operator, and special character actually means. We answer questions like:

- Why do some things use () while others use [] or {}?
- What's the difference between = and == and ===?
- Why does my code break when I use & instead of &&?
- What does -> do differently from . or ::?
- When do I need semicolons and when don't I?

These aren't trivial questions. **They're the foundation of everything else in programming.** Once you understand the symbols, the actual programming concepts become dramatically easier to learn.

The Truth About Learning Programming

Here's what most people won't tell you: **the hardest part of learning programming isn't the logic—it's the syntax.**

You can understand perfectly how an if-statement should work (if the condition is true, do this; otherwise, do that). But if you don't know that >= means "greater than or equal to" or that && means "and," you can't write the actual code.

This book solves that problem. It teaches you the language before teaching you the grammar. Once you know the symbols, everything else becomes dramatically easier.

Why I Wrote This Book

I wrote this book because I was you. I started learning programming from free online courses, got completely lost in the sea of symbols, and almost gave up. I'd see code like this:

javascript

```
const result = data?.filter(x => x.value > 10)
        .map(x => x.name)
        .join(', ') ?? 'No results';
```

And think, "I'll never understand this. Programming isn't for me."

48

But then I realized something: **I didn't need to be smarter. I just needed someone to explain what the symbols meant.**

Once I understood that ?. prevents errors on null values, that => creates functions, that . chains methods together, and that ?? provides defaults, the code became perfectly readable. It wasn't magic—it was just symbols I hadn't learned yet.

That's why this book exists. To give you the foundation that I wish I'd had when I started.

Your Journey Starts Here

Programming isn't a secret club for geniuses. It's a skill that anyone can learn, **if they start with the right foundation.**

This book gives you that foundation. Once you understand the symbols, you'll be able to:

- Follow any programming tutorial with confidence
- Read code examples and understand what they do
- Write your own code without constantly googling syntax
- Learn new languages quickly by recognizing familiar patterns
- Think like a programmer, not just copy like a student

The symbols are the key. Everything else in programming builds on this foundation.

Turn the page and start your programming journey the right way—from the ground up, with complete understanding.

Who This Book Is For

This book is written for anyone who has ever stared at a line of code and thought: *"What does that symbol even mean?"*

It's designed to help beginners and self-learners understand programming from the ground up—by focusing on **symbols**, the true building blocks of code.

This book is for you if:

- You're completely new to programming and want a strong, practical foundation.
- You've taken online courses but still feel lost when unfamiliar symbols appear.
- You can copy code examples that work but don't fully understand *why* they work.
- You're learning multiple programming languages and get confused by their differences.
- You want to read and understand other people's code with confidence.
- You're tired of memorizing syntax without grasping the logic behind it.

This book is especially valuable if:

- You learn best when you understand *why* something works, not just *how*.
- You've felt intimidated by programming because it seems like a secret language.
- You want to move beyond tutorials and start creating your own projects.
- You prefer clear, step-by-step explanations before moving on to the next concept.

How This Book Works

This book is organized by **symbol categories**, progressing from the most fundamental to the most advanced. Each section builds upon the previous one, so by the end, you'll not only recognize symbols—you'll understand their logic, meaning, and purpose in real code.

FOUNDATION SYMBOLS (Chapters 1–6)

- **Brackets and Parentheses:** The structural backbone that shapes your code.
- **Comparison Operators:** How programs make logical decisions.
- **Arithmetic Operators:** Performing mathematical operations and calculations.
- **Assignment Operators:** Storing and managing variable values.
- **Logical Operators:** Connecting multiple conditions to form complex logic.
- **Bitwise Operators:** Understanding how computers process data in binary form.

ADVANCED SYMBOLS (Chapters 7–10)

- **Special Characters:** Punctuation and formatting that influence readability and structure.
- **Arrow Operators:** Compact and modern syntax used in languages like JavaScript and C++.
- **Language-Specific Symbols:** Unique operators that define the "personality" of each programming language.
- **Complete Reference Guide:** A quick lookup section for reviewing and troubleshooting code.

Each Chapter Includes

- **Plain-English explanations** that cut through technical jargon.

- **Visual examples** showing exactly how each symbol works.
- **Real code samples** from multiple languages (Python, JavaScript, Java, and more).
- **Common beginner mistakes**—and how to avoid them.
- **Hands-on exercises** to strengthen understanding through practice.
- **Cross-language comparisons** that help you think like a programmer, not just memorize syntax.

A Note to the Reader

As you begin this journey, remember that learning to read symbols is like learning to read music—each one has a rhythm, a tone, and a purpose.

Take your time, explore each chapter, and celebrate every "aha!" moment along the way. By the end of this book, you won't just be reading code—you'll be *understanding its language* and thinking like a true programmer.

How to Use This Book

For Complete Beginners: Read this book before starting any programming course. It will give you the foundation that courses assume you already have. When an instructor uses a symbol, you'll know exactly what it means and why.

For Self-Learners: Keep this book next to you while learning from online tutorials. When you see an unfamiliar symbol, look it up in this book. Understanding the symbol will help the rest of the tutorial make sense.

For Career Changers: Use this book to build confidence. You don't need a computer science degree to understand programming symbols. This book explains everything from scratch, assuming no prior knowledge.

For Multi-Language Learners: Read through the book once to understand the concepts, then use it as a reference. The cross-language comparisons will help you see that the same ideas appear in different languages with slightly different symbols.

For Visual Learners: This book is designed for you. Every concept is demonstrated with concrete examples and visual code blocks. You'll see exactly what each symbol does, not just read abstract descriptions. [Note: Add diagrams here—e.g., a flowchart for parentheses usage.]

What You'll Gain From This Book

Immediate Benefits:

· Confidence to read and understand any code example you encounter

· The ability to fix your own syntax errors instead of random guessing

· Understanding of why code works, not just that it works

· Freedom from copy-paste programming—you'll understand what you're typing

Long-Term Benefits:

· Faster learning of new programming languages (the symbols are mostly the same)

· Ability to switch between languages without confusion

· Foundation for advanced programming concepts

· Confidence to contribute to real projects and understand professional code

Career Benefits:

· Stand out in job interviews by demonstrating deep understanding

· Debug code more efficiently (most bugs are simple symbol mistakes)

· Write cleaner, more readable code that follows professional standards
· Communicate more effectively with other programmers
The Truth About Learning Programming
Here's what most people won't tell you: the hardest part of learning programming isn't the logic—it's the syntax.
You can understand perfectly how an if-statement should work (if the condition is true, do this; otherwise, do that), but if you don't know that == checks equality while = assigns a value, your code won't run. This book tackles that head-on, giving you the tools to master syntax before diving into bigger ideas.

Why Symbols Matter More Than You Think

Comprehensive Symbol Reference
Below is a complete reference of the programming symbols covered in this book, grouped by category, with their meanings, examples in multiple languages (Python, JavaScript, Java), and common uses. Each entry is designed to be concise yet informative, so you can quickly find what you need.

Grouping Symbols
These symbols organize your code, acting like containers for logic and data.
- Parentheses ()
  - Purpose: Group expressions, define function calls, or set precedence in calculations.
  - Example (Python):
    result = (2 + 3) * 4 # Parentheses ensure 2 + 3 is calculated first
    print(result) # Outputs: 20
  - Example (JavaScript):
    function greet(name) {
       return `Hello, ${name}`;
    }
    console.log(greet("Alice")); # Outputs: Hello, Alice
  - Common Use: Ensuring order of operations, passing arguments to functions.
- Square Brackets []
  - Purpose: Access array/list elements or define arrays.
  - Example (Java):
    int[] numbers = {1, 2, 3};
    System.out.println(numbers[0]); # Outputs: 1
  - Common Use: Storing and retrieving list or array data.
- Curly Braces {}
  - Purpose: Define code blocks, objects, or dictionaries.
  - Example (JavaScript):
    let user = { name: "Bob", age: 25 };
    console.log(user.name); # Outputs: Bob
  - Common Use: Grouping statements in loops/functions or defining key-value pairs.

Comparison Symbols
These symbols make decisions by comparing values.
- Equal to ==
  - Purpose: Checks if two values are equal.
  - Example (Python):
    if 5 == "5": # Type coercion in some languages
        print("Equal")
  - Common Use: Testing conditions in if statements.
- Not Equal !=
  - Purpose: Checks if two values are not equal.
  - Example (Java):
    if (score != 100) {
        System.out.println("Try again!");
    }
- Greater/Less Than >, <, >=, <=
  - Purpose: Compare numerical values.
  - Example (JavaScript):
    let age = 18;
    if (age >= 18) {
        console.log("Adult");
    }

Math Symbols
These handle calculations and numerical operations.
- Basic Operations +, -, *, /
  - Purpose: Perform addition, subtraction, multiplication, division.
  - Example (Python):
    total = 10 * 2 / 5 # Outputs: 4.0
- Modulus %, Exponentiation **, Floor Division //
  - Purpose: Find remainders, raise to powers, or divide with integer results.
  - Example (Python):
    print(10 % 3) # Outputs: 1
    print(2 ** 3) # Outputs: 8
    print(10 // 3) # Outputs: 3
- Increment/Decrement ++, --, Compound Assignments +=, -=, *=
  - Purpose: Modify values efficiently.
  - Example (Java):
    int x = 5;
    x += 3; # x is now 8

Logic Symbols
These connect conditions to build complex logic.
- Logical AND &&, OR ||, NOT !
  - Purpose: Combine or invert conditions.
  - Example (JavaScript):
    if (age >= 18 && hasLicense) {

```
    console.log("Can drive");
  }
```
- Bitwise AND &, OR |, XOR ^, NOT ~
  - Purpose: Manipulate bits directly.
  - Example (Python):
```
permissions = 5 & 3  # Binary: 0101 & 0011 = 0001
print(permissions)  # Outputs: 1
```
- Shift Operators <<, >>
  - Purpose: Shift bits left or right for fast multiplication/division.
  - Example (Java):
```
int x = 8 >> 1;  # Binary: 1000 >> 1 = 0100 (4 in decimal)
```

Assignment Symbols
These store or update values.
- Basic Assignment =
  - Purpose: Assign a value to a variable.
  - Example (Python):
```
x = 10
```
- Compound Assignments +=, -=, *=
  - Purpose: Combine operations with assignment.
  - Example (JavaScript):
```
let score = 100;
score *= 2;  # score is now 200
```

Special Symbols
These handle unique tasks like accessing objects or ending statements.
- Comments //, #, /* */
  - Purpose: Add notes or disable code.
  - Example (Java):
```
// This is a single-line comment
/* This is
   a multi-line comment */
```
- Object Access ., ->
  - Purpose: Access properties or methods.
  - Example (JavaScript):
```
let user = { name: "Alice" };
console.log(user.name);  # Outputs: Alice
```
- Statement Terminator ;
  - Purpose: End statements in languages like Java and JavaScript.
  - Example (Java):
```
System.out.println("Hello");  # Semicolon required
```

Troubleshooting Common Symbol Errors
Even with a solid understanding of symbols, errors happen. Below are the most common issues beginners face, based on my own coding mishaps, along with how to fix them.

1. Mismatched Grouping Symbols
   - Error: "SyntaxError: unexpected EOF" or "missing }"
   - Cause: Unclosed parentheses, brackets, or braces.
   - Example (JavaScript):

```
function example() {
    if (true) {
        console.log("Hi");
    // Missing closing }
}
```

   - Fix: Count your opening and closing symbols. Use an editor like Visual Studio Code that highlights matches.

```
function example() {
    if (true) {
        console.log("Hi");
    } // Added closing }
```

2. Confusing = with ==
   - Error: Code runs but produces wrong results.
   - Cause: Using assignment (=) instead of comparison (==) in conditions.
   - Example (Python):

```
if x = 5:  # Assigns 5 to x, causing an error
    print("This won't work")
```

   - Fix: Use == for comparisons.

```
if x == 5:
    print("This works")
```

3. Mixing Logical and Bitwise Operators
   - Error: Unexpected logic in conditions.
   - Cause: Using & instead of && or | instead of ||.
   - Example (Java):

```
if (age > 18 & hasLicense) {  // Bitwise, not logical
    System.out.println("Wrong operator");
}
```

   - Fix: Use logical operators for conditions.

```
if (age > 18 && hasLicense) {
    System.out.println("Correct");
}
```

4. Forgetting Semicolons in Java/JavaScript
   - Error: "SyntaxError: missing ; before statement"
   - Cause: Omitting ; in languages that require it.
   - Example (Java):

```
int x = 5  // Missing semicolon
System.out.println(x);
```

   - Fix: Add semicolons where needed.

```
int x = 5;
System.out.println(x);
```

5. Incorrect Bitwise Operations
  - Error: Wrong numerical results.
  - Cause: Misunderstanding bitwise operators like ^ or <<.
  - Example (Python):
  x = 5 ^ 2 # Expected toggle, but unclear binary result
  print(x) # Outputs: 7 (binary 0101 ^ 0010 = 0111)
  - Fix: Visualize binary:
  5: 0101
  2: 0010
  ^: 0111 (7)
  Test with small numbers and print intermediate results.

6. Language-Specific Symbol Differences
  - Error: Code works in one language but fails in another.
  - Cause: Symbols like ** (Python) vs Math.pow (Java) differ.
  - Example:
  x = 2 ** 3 # Works in Python
  int x = 2 ** 3; # Fails in Java
  - Fix: Use language-appropriate symbols.
  int x = (int) Math.pow(2, 3); # Correct for Java

Visual Cheat Sheet
To make symbols stick, here's a quick visual reference:

Category	Symbols	Example (Python)
Grouping	(), [], {}	x = [1, 2, 3]
Comparison	==, !=, >, <, >=, <=	if x == 5:
Math	+, -, *, /, %, **, //	x = 10 % 3
Logic	&&, ||, !, &, |, ^, ~	if x > 0 and y > 0:
Assignment	=, +=, -=, *=, /=	x += 5
Special	., ;, #, //, ->	print(x.y)

[Note: Add a printable version of this table with color coding for categories.]

Practice Opportunities
Reinforce your skills with these exercises:
1. Symbol Hunt: Write a program in Python that uses at least one symbol from each category (grouping, comparison, math, logic, assignment, special). Print a message based on a condition.
2. Debug Challenge: Take this buggy code and fix it:
```
let x = 10
if (x = 5) {
    console.log("This won't work");
```

}
3. Bitwise Fix: Correct this code to check if 4 is set in permissions = 7:
  permissions = 7
  if permissions & 4:
      print("Write permission")
Test these in Replit, CodePen, or Visual Studio Code.

Final Tips for Success
- Use Visual Editors: Tools like Visual Studio Code highlight matching brackets and flag missing semicolons.
- Test Incrementally: Write small chunks of code and test often to catch symbol errors early.
- Google Errors: Search error messages with the language name (e.g., "JavaScript unexpected token }") for specific fixes.
- Practice Daily: Spend 5 minutes writing code with different symbols to build muscle memory.

Conclusion
You've now unlocked the hidden language of programming symbols! From organizing code with () and {} to making decisions with == and &&, you have the tools to read, write, and debug code confidently. This reference and troubleshooting guide is your companion for the journey ahead. Keep experimenting, stay curious, and don't be afraid to make mistakes— every error is a step toward mastery. Happy coding!

🖥️ *Hardware Family — The Machines That Run Your Code*

When you write code, it's just text — it doesn't "do" anything until a **computer's hardware** runs it. Hardware is the **physical part** of a computer or device. It takes the instructions you write and turns them into actions: calculations, graphics, artificial intelligence, and more.

Think of it like this:

- Your **code** is a recipe.
- The **hardware** is the kitchen — it has the tools to cook your recipe and produce the final dish (the running program).

Every type of hardware in this family has a special job:

- The **CPU (Central Processing Unit)** is the "brain," handling most general tasks.
- The **GPU (Graphics Processing Unit)** is a powerful multitasker for graphics and AI.
- The **TPU (Tensor Processing Unit)** is built just for machine learning.
- The **NPU (Neural Processing Unit)** powers smart features on your phone or laptop.
- Other chips like **ASICs** and **FPGAs** are designed for very specific jobs.
- **Memory (RAM)** and **Storage (HDD/SSD)** keep your code and data ready for action.

Knowing this family isn't about becoming a hardware engineer — it's about **understanding where your code runs**. As you grow from beginner to advanced coder, this knowledge helps you choose the right tools, understand performance limits, and work with new tech like AI.

🖥️ *Hardware Family — The Machines Behind Your Code*

When you write code, it doesn't do anything by itself. It needs **hardware** — the physical parts of a computer — to run and make your programs work. Understanding the main hardware families will help you see **where your code lives and how it runs**.

CPU (Central Processing Unit)

- **What it is:** The "brain" of the computer.
- **What it does:** Handles most general tasks — reading your code, doing calculations, and controlling other parts of the computer.
- **Responsibilities:** Runs instructions one after another, great for apps, web browsing, office work, and small programs.
- **For beginners:** Almost every piece of code you write will run on a CPU by default.

GPU (Graphics Processing Unit)

- **What it is:** A super-fast calculator with thousands of tiny workers.
- **What it does:** Originally built to draw graphics for games and videos; now used for heavy math tasks like AI and 3D rendering.
- **Responsibilities:** Performs many simple calculations at the same time — perfect for **AI training, deep learning, video editing, and gaming**.
- **For beginners:** You don't need a GPU for basic coding, but it's important if you want to learn **AI or data science** later.

TPU (Tensor Processing Unit)

- **What it is:** A special chip designed by Google for **machine learning and AI**.
- **What it does:** Optimized for running neural networks quickly and efficiently.
- **Responsibilities:** Speeds up AI tasks like training chatbots, image recognition, and language models.
- **For beginners:** You might use TPUs later if you train big AI models in Google Cloud.

NPU (Neural Processing Unit)

- **What it is:** A chip made for **AI tasks on everyday devices** like phones and laptops.
- **What it does:** Handles tasks such as voice recognition, photo enhancements, and smart assistants without draining battery.
- **Responsibilities:** Runs AI features directly on your device — think Siri, Face ID, or real-time photo filters.

- **For beginners:** You use NPUs without knowing it when your phone does AI tasks.

ASIC (Application-Specific Integrated Circuit)

- **What it is:** A chip designed to do **one job really well**.
- **What it does:** Runs a single specialized task like cryptocurrency mining, network routing, or image processing.
- **Responsibilities:** Extremely fast for its job, but can't be easily repurposed.
- **For beginners:** You probably won't buy one unless you're doing specialized hardware projects.

FPGA (Field-Programmable Gate Array)

- **What it is:** A chip you can **reprogram to act like any other chip**.
- **What it does:** Lets engineers design and test new hardware logic without building a new chip from scratch.
- **Responsibilities:** Used in research, hardware prototypes, and some high-speed financial trading.
- **For beginners:** Rarely needed, but interesting for hardware engineering students.

Memory (RAM)

- **What it is:** Short-term memory of your computer.
- **What it does:** Temporarily stores data and instructions while your program is running.
- **Responsibilities:** The more RAM you have, the more apps and large data sets you can handle at once.
- **For beginners:** Needed for running code smoothly — especially if your program uses big files.

Storage (HDD & SSD)

- **What it is:** Long-term memory of your computer.
- **What it does:** Saves your code, projects, videos, and files even when the power is off.
- **Responsibilities:** SSDs are faster than HDDs; faster storage means your programs load quicker.
- **For beginners:** Where your coding files and projects are saved.

Why This Matters for Coders

- Your **CPU** runs most programs you'll write.
- **GPUs** and **TPUs** are needed for **AI, data science, and big graphics tasks**.
- **NPUs** make your phone and laptop smarter.
- **RAM & Storage** help your code run faster and store data.

You don't have to buy all this hardware now — but knowing what powers your code helps you **choose the right tools** as you grow.

Code Breaker Families

When you start learning to code, the hardest part isn't always the logic — it's the strange symbols.
Brackets, slashes, curly things, dots, equals signs… they all look mysterious at first.
This book was built to **break the code** and make those symbols simple.

To help you, I've organized the most important coding symbols and rules into **friendly "families."**
Each family groups related symbols, explains them in plain English, and shows tiny examples so you can **see them in action**.

You don't need to memorize everything.
Use these families as your **quick guide and safety net** whenever you get stuck.

How to Use the Families

1. **Skim first** — glance through to see what symbols exist and what they do.
2. **Flip back often** — when you hit an error or confusing code, check the right family.
3. **Try the examples** — copy them into your code editor and see what happens.
4. **Build your confidence** — as you learn, these symbols will feel natural.

The Families You'll Meet

- **Quick Symbol Cheat Sheets** – fast lookup tables for HTML, CSS, Python, and JavaScript.
- **The Brackets & Braces Family** – `()`, `[]`, `{}`, `<>` and how they hold your code together.
- **The Operators Family** – `=`, `==`, `===`, `+`, `-`, `*`, `/` and other action heroes.
- **The Comment Family** – how to leave notes in code using `#`, `//`, `/* */`, and `<!-- -->`.
- **The String Family** – quotes, escape characters, backticks, and joining text.
- **The Path & URL Family** – `./`, `../`, `/`, and how to tell your code where files live.
- **The Safety & Error Family** – `try/except`, `try/catch`, colons, and semicolons.
- **The Syntax Survival Kit** – fixes for the most common beginner mistakes.
- **The Numbers & Math Symbols Family** – `+`, `-`, `*`, `/`, `%`, and other math tools.
- **The Comparison & Logic Family** – `>`, `<`, `>=`, `==`, `and`, `or`, `!`.
- **The Variables & Naming Family** – how to name things and use `_`, camelCase, `$`.
- **The Functions & Calls Family** – `def`, `function`, `()`, `return`, and arrow functions.
- **The File & Folder Family** – paths, file extensions, and keeping your project organized.
- **The Web & Link Family** – anchors, IDs, classes, and linking pages together.

Beginner Promise:

You don't have to be a tech expert to understand this.

Each family explains **what the symbol means, why it matters, and shows a simple example**.

By the end, you'll read code with confidence — and write your own without fear of the symbols.

The Syntax Family — The Rulebook Behind All Other Families

- **What It Is:**
  Syntax is the *grammar* of coding. It tells you where symbols can appear and how to combine them so the computer understands your code.
- **Key Elements of Syntax:**
    - **Keywords:** `if`, `for`, `while`, `class`, `return`
    - **Indentation & Spacing:** especially important in Python.
    - **Punctuation:** commas, semicolons, colons, and parentheses.
    - **Order:** where variables, functions, and braces appear.
- **Why It's Not a Regular Family:**
  Unlike Math or Bitwise operators, syntax is not a list of characters. It's a *system of rules* that controls how all the other families fit together.
- **Examples in Different Languages:**

```
# Python Syntax: indentation matters
if age >= 18:
    print("Adult")
// JavaScript Syntax: braces + semicolons
if (age >= 18) {
    console.log("Adult");
}
// Java Syntax: strict types and semicolons
int age = 18;
if (age >= 18) {
    System.out.println("Adult");
}
```

🔍 *Welcome to the Code Breaker Quick Reference Family*

Before we dive deep into the chapters, meet the newest member of the **Code Breaker family** — the **Quick Symbol Cheat Sheets**.

Think of these pages as your **first-aid kit for confusing code symbols**. Whenever you get lost while reading a tutorial, debugging your own project, or just wondering *"What does this curly thing mean?"* you can flip here and find the answer fast.

Each cheat sheet is grouped by language — HTML, CSS, Python, and JavaScript — so you can jump right to the one you need. You'll see the symbol, its name, what it does, and a simple example.

🧩 Quick Symbol Cheat Sheet — HTML

Symbol	Name	Meaning / Use	Example
< >	Angle Brackets	Wrap HTML tags to mark elements.	`<p>Hello</p>`
</ >	Closing Tag Slash	Closes an HTML element.	`</h1>`
=	Attribute Assignment	Gives a value to an attribute.	`<img src="cat.jpg">`
" "	Quotation Marks	Surround attribute values.	`<a href="home.html">`
/ (inside < />)	Self-Closing Indicator	Ends empty tags.	` `
<!-- -->	Comment	Adds notes that don't show on the page.	`<!-- This is a comment -->`

Tip: In HTML, tags are like containers. Angle brackets hold the tag name, attributes add details, and the slash / tells the browser "this is the end."

🖥️ *The Coding Platforms Family*

Coding platforms are the **places where programmers write, test, and share their code.** They provide the tools, environments, and support needed to turn ideas into working software. Just like an artist needs a studio, a programmer needs a coding platform.

1. Code Editors (Lightweight & Flexible)

- **What They Are:** Simple, fast programs for writing code with helpful features like color highlighting and auto-complete.
- **Best For:** Beginners and quick projects.
- **Popular Tools:**
  - **Visual Studio Code (VS Code)** – free, highly customizable

o **Sublime Text** – lightweight and fast

o **Atom** – open source, easy for web projects

Example:

```
my_project/
 └ hello_world.py  ← file opened in VS Code
```

2. Integrated Development Environments (IDEs)

- **What They Are:** All-in-one workspaces with a code editor, debugger, compiler, and project manager.
- **Best For:** Larger projects and professional software development.
- **Popular Tools:**
  - o **PyCharm** – Python
  - o **IntelliJ IDEA** – Java & Kotlin
  - o **Eclipse** – Java/C++
  - o **Visual Studio** – C#, C++, .NET

Example:

```
// Running Java code inside IntelliJ IDEA
public class Main {
    public static void main(String[] args) {
        System.out.println("Hello, World");
    }
}
```

3. Online Coding Platforms (Browser-Based)

- **What They Are:** Websites that let you code and run programs directly in the browser — no installation required.
- **Best For:** Learning, quick sharing, and experimenting anywhere.
- **Popular Tools:**
  - o **Replit** – supports many languages
  - o **CodePen** – front-end web (HTML/CSS/JS)
  - o **JSFiddle** – JavaScript demos
  - o **Google Colab** – Python & AI experiments

Example:

```
# Code typed and run in Google Colab
print("Hello from the cloud!")
```

4. Cloud Development Environments

- **What They Are:** Full IDEs hosted in the cloud so you can work from any device.
- **Best For:** Teams, remote work, and advanced projects.
- **Popular Tools:**
    - **GitHub Codespaces** – cloud-based VS Code
    - **AWS Cloud9** – online coding with collaboration
    - **Gitpod** – instant dev environments linked to GitHub

Example:
A developer opens **GitHub Codespaces** in the browser, edits files, and runs code without installing anything locally.

5. Mobile Coding Apps

- **What They Are:** Apps for phones and tablets that let you practice or write small scripts on the go.
- **Best For:** Students, learning, and quick fixes.
- **Popular Tools:**
    - **Pydroid 3** – Python on Android
    - **Koder** – iOS code editor
    - **Juno** – Jupyter notebooks on iPad

6. Version Control & Collaboration Platforms

- **What They Are:** Platforms that store your code online and allow multiple developers to work together safely.
- **Best For:** Sharing, team projects, and open source development.
- **Popular Tools:**
    - **GitHub** – most popular for open source and private repos
    - **GitLab** – for teams and enterprise use
    - **Bitbucket** – integrates with Jira & Trello

☑ Quick Comparison Table

Platform Type	Examples	Best For
Code Editors	VS Code, Sublime Text	Quick coding, small projects
IDEs	PyCharm, IntelliJ, Eclipse	Professional, large software projects
Online Platforms	Replit, CodePen, Colab	Learning, sharing, quick experiments
Cloud Dev Environments	GitHub Codespaces, Cloud9	Remote work, collaboration
Mobile Coding Apps	Pydroid, Juno	Practice and coding on the go
Version Control	GitHub, GitLab, Bitbucket	Team collaboration, safe code storage

🚀 Takeaway

The **Coding Platforms Family** shows that programming isn't tied to one place.

- Beginners often start with **online platforms** or **simple editors**.
- Professionals use **IDEs and cloud environments**.
- Teams rely on **version control platforms** to collaborate and share work safely.

Tip: Try a simple online platform (like Replit or CodePen) to practice symbols and logic, then move to a full IDE when your projects grow.

🧩 The Brackets & Braces Family

Before you start writing real programs, you'll notice that code is full of different kinds of **brackets**. Each has a special job. Think of them as **family members with different personalities**:

Parentheses ( ) — The Callers

Parentheses wrap things you **call** or **group**.

- In Python and JavaScript they are used to **run a function**:
- `print("Hello")`

  Here `()` tells Python to run the `print` function.

- In math expressions they group things: `(2 + 3) * 4`.

Square Brackets [ ] — The Organizers

Square brackets hold **lists or arrays** (collections of items) and let you **pick** one item by its position.

```
colors = ["red", "blue", "green"]
print(colors[0])   # prints "red"
```

The `[0]` means "give me the first thing."

Curly Braces { } — The Block Builders

Curly braces group code or store **key–value pairs**.

- In JavaScript, they create **objects**:
- `let user = { name: "Susie", age: 30 };`
- In CSS, they hold style rules:
- `h1 { color: blue; }`

Angle Brackets < > — The Taggers

Angle brackets mark **HTML tags** — the building blocks of a web page.

```
<h1>Welcome</h1>
<p>This is a paragraph.</p>
```

They tell the browser what each part of the page is.

💡 **Tip for Beginners:**
If you ever get a "missing bracket" error, count each one and make sure they **open and close in pairs**. Every (must have a), every { needs a }, and so on.

⚙️ *The Operators Family*

Every programming language uses **operators** — special symbols that **do something** with values. Think of them as the **action heroes** of code: they compare things, join things, or perform math.

Assignment Operator (=) — The Setter

When you use a single equals sign, you're **putting a value into a variable**.
It's like saying: "Hey computer, remember this!"

```
name = "Susie"      # Save the word Susie into the variable name
age = 30            # Save the number 30 into the variable age
```

Equality Operator (==) — The Question Asker

Two equals signs ask: **"Are these the same?"**
It doesn't set a value — it compares two things and answers **True** or **False**.

```
5 == 5      # True
5 == 8      # False
```

Strict Equality (===) — The Double Checker (JavaScript)

In JavaScript, === checks **both value and type**.
5 === "5" is False because one is a number and the other is a string.

```
5 === 5      // true
5 === "5"    // false
```

Not Equal (!= or !==) — The Opposite Asker

This asks: **"Are these different?"**

- Python uses `!=`
- JavaScript uses `!==` for strict checking

```
3 != 4    # True (they are different)
3 != 3    # False (they are the same)
3 !== "3"   // true (different type)
3 !== 3     // false
```

Math Operators (+ - * / %) — The Number Crunchers

- `+` Add
- `-` Subtract
- `*` Multiply
- `/` Divide
- `%` Remainder (what's left after division)

```
5 + 2  # 7
5 - 2  # 3
5 * 2  # 10
5 / 2  # 2.5
5 % 2  # 1  (remainder when 5 is divided by 2)
```

Logical Operators (and, or, not in Python / &&, ||, ! in JS) — The Decision Makers

These combine **True/False** values to make decisions.

```
age = 20
if age > 18 and age < 30:
    print("Young adult")
let loggedIn = true;
let admin = false;
if (loggedIn && admin) {
    console.log("Welcome Admin");
}
```

💡 Beginner Tip:
Remember:

- `=` is for **setting** a value.
- `==` and `===` are for **asking** if values are equal.
- Logic symbols help your program **decide** what to do next.

💬 The Comment Family

When you're writing code, you'll often want to leave yourself **notes** or explain what a section of code does. These notes are called **comments**.
Comments are **ignored by the computer** — they're only for humans to read. They help you remember what you were doing and make your code easier for others to understand.

— Python's Note Marker

In Python, any line that starts with a # is a comment.

```
# This prints a welcome message
print("Hello, world!")
```

Everything after the # on that line is just a note.

// — Single-Line Comment in JavaScript, CSS

Use // in JavaScript to write a quick one-line note.

```
// This shows a message in the console
console.log("Hello!");
```

/* ... */ — Multi-Line Comment in JavaScript & CSS

If your note is longer than one line, wrap it between /* and */.

```
/* This is a CSS style
   that makes the heading red */
h1 {
  color: red;
}
/* This function adds two numbers
   and returns the result */
function add(a, b) {
  return a + b;
}
```

<!-- ... --> — HTML Comment

In HTML, comments look like this:

```
<!-- This section is the main heading -->
<h1>Welcome to my page</h1>
```

They won't show up on the webpage, but they help you keep track of what's where.

💡 Beginner Tip:

Comments are **your friend**. Use them to explain why you wrote code a certain way. It will save you (and anyone reading your code) a lot of confusion later.

🎩 *The String Family*

When you work with code, you'll often need to handle **text** — words, sentences, or even whole paragraphs. In programming, text is called a **string**.
Strings let your programs display messages, hold user names, show website text, and more. Understanding the symbols around strings will save you many errors.

Quotation Marks " " and ' ' — String Wrappers

You wrap text inside quotes to tell the computer "this is a string."

```
name = "Susie"
greeting = 'Hello, world!'
```

- In Python and JavaScript, you can use either **single** (' ') or **double** (" ").
- Just make sure you **start and end with the same type**.

Backtick ` ` — Template Strings (JavaScript)

Backticks let you make strings that can **insert variables** easily.

```
let name = "Susie";
console.log(`Hello, ${name}!`); // Prints: Hello, Susie!
```

Anything inside ${ } will be replaced with the variable's value.

Escape Character \ — Special Helpers Inside Strings

Sometimes you need special characters in your string. The backslash \ tells the computer the next symbol is special.

- \n = new line
- \t = tab space
- \\ = a backslash itself

```
print("Hello\nWorld")  # prints on two lines
```

String Concatenation + — Joining Text Together

The plus sign can join strings.

```
let first = "Hello";
```

```
let second = "World";
console.log(first + " " + second); // Hello World
```

In Python you can also use +, but f-strings (like `f"Hello {name}"`) are more modern.

Triple Quotes """ """ — Multi-Line Strings (Python)

When your text needs to stretch across several lines, Python lets you use triple quotes.

```
story = """Once upon a time,
there was a coder learning symbols."""
print(story)
```

💡 Beginner Tip:
Strings are just text, but the symbols around them (", ', `, \) change how they work.
If you get an error, check if your quotes match — many beginner mistakes come from a missing or mismatched quote.

🌐 *The Path & URL Family*

When you build websites or work with files, you'll see a lot of **slashes and dots** in paths and web addresses.
These little symbols tell your code **where to look** for a file or a page.
Think of them as a **map** to your content.

Forward Slash / — Folder Separator

In file paths and URLs, a forward slash separates folders.

```
<img src="images/photo.jpg">
```

Here the browser goes into the **images** folder to find `photo.jpg`.

Double Dot ../ — Go Up a Folder

Two dots mean "go up one level in the folder structure."

```
../index.html
```

If you're inside a subfolder, `../` jumps up to the parent folder.

Single Dot ./ — Current Folder

A single dot means "start from the current folder."

```
./script.js
```

This tells the browser to look for `script.js` right where the HTML file is.

Protocol (`http://` or `https://`) — How to Access a Site

Web addresses start with a protocol.

- `http://` = HyperText Transfer Protocol (basic)
- `https://` = Secure version (encrypted)

```
<a href="https://example.com">Visit Example</a>
```

Most modern sites use **https** for safety.

Hash # — Jump to a Section

A hash links to a specific part of a page.

```
<a href="#contact">Contact Us</a>
```

If the page has an element with `id="contact"`, the browser will scroll straight to it.

🔍 Beginner Tip:
When your code can't find a file or shows a "404 not found" error, check your path:

- Are you using the right number of `../` to go up?
- Did you spell the folder/file names exactly (case matters on some systems)?
- Are you linking with `https://` if it's a full web address?

🛡 The Safety & Error Family

Even the best code will sometimes fail — a typo, a missing symbol, or a situation the program didn't expect.
These special words and symbols help you **handle errors safely** so your program doesn't crash and burn.

`try:` — Safe Test Zone (Python)

`try:` tells Python:
"Try to run this code. If it breaks, don't crash — I'll handle it."

```
try:
    number = int("abc")  # this will cause an error
except:
    print("Something went wrong!")
```

If an error happens, the program jumps to the `except` part instead of stopping.

`except` — The Catcher (Python)

Used with `try:`.
`except` catches the error and lets you decide what to do next.

```python
try:
    print(10 / 0)  # dividing by zero causes an error
except:
    print("You can't divide by zero!")
```

`throw` / `catch` — Error Tossing in JavaScript

JavaScript uses `throw` to signal an error and `catch` to handle it.

```javascript
try {
  throw new Error("Something broke!");
} catch (e) {
  console.log(e.message);
}
```

If something goes wrong inside the `try` block, `catch` grabs the error.

Semicolon `;` — Statement Stopper

In JavaScript and CSS, a semicolon ends a statement or rule.
It helps avoid confusion when the computer reads your code.

```
let x = 5;
let y = 10;
console.log(x + y);
p {
  color: blue;
  font-size: 16px;
}
```

Colon `:` — Block Starter

In Python, a colon starts an **indented block** — code that belongs to something (like a loop or condition).

```python
if 5 > 3:
    print("Five is greater!")
```

💡 **Beginner Tip:**
Think of `try`/`except` (Python) or `try`/`catch` (JavaScript) as **safety nets**.
They keep your program from crashing when something unexpected happens.
Colons `:` in Python and semicolons `;` in JavaScript/CSS are like **punctuation** — tiny but important!

⏱ *The Syntax Survival Kit*

Every beginner eventually runs into **syntax errors** — the computer saying "I don't understand this code."
These problems usually come from **missing or mismatched symbols**.
Think of this section as a **survival kit** to help you fix the most common beginner mistakes fast.

Mismatched Quotes

Strings must start and end with the **same kind of quote**.

```
name = "Susie'     # ✖ wrong: starts with " but ends with '
name = "Susie"     # ▨ correct
```

Unclosed Brackets or Braces

Every opening ([{ must have a closing)] }.

```
if (score > 10 {        // ✖ missing )
if (score > 10) {       // ▨ correct
```

Tip: When an error says *"unexpected end of input"* or *"missing)"*, check your pairs.

Wrong Indentation (Python)

Python cares about spaces at the start of a line.
Each block (after a :) must be indented the same amount.

```
if True:
print("Hello")    # ✖ not indented
    print("Hello")   # ▨ correct
```

Forgotten Semicolons (JavaScript & CSS)

While modern JavaScript can sometimes skip semicolons, forgetting them can break code.
CSS **must** end each rule with a semicolon.

```
p {
  color: red      /* ✖ missing ; */
  font-size: 16px;
}

p {
```

```
  color: red;    /* ☑ correct */
  font-size: 16px;
}
```

Wrong Case (File Paths & HTML IDs)

On many servers, `style.css` and `Style.css` are **not the same**.
Keep names consistent.

```
<link rel="stylesheet" href="style.css">  <!-- must match the file exactly
-->
```

Extra or Missing Colons

In Python, every `if`, `for`, `while`, or `def` needs a colon at the end.

```
if x > 5          # ✖ missing colon
if x > 5:         # ☑ correct
```

💡 Beginner Tip:
If your program shows an error message, **read it slowly** — it often points to the line and even the symbol causing the issue.
Most early bugs come from **tiny typos** like these, so don't panic — just check your symbols.

▦ The Numbers & Math Symbols Family

Every program uses **numbers**.
These symbols let your code add, subtract, and do more advanced math.
Think of them as the **calculator tools** of coding.

Plus + — Addition & Joining

- **Math:** Adds numbers.
- `3 + 2  # 5`
- **Strings:** Joins (concatenates) text in many languages.
- `"Hello" + " World" // Hello World`

Minus - — Subtraction & Negatives

- **Math:** Subtracts one number from another.
- `10 - 3  # 7`
- **Negative numbers:**
- `-5  # negative five`

Asterisk * — Multiplication & Repetition

- **Math:** Multiply numbers.
- `4 * 5  # 20`
- **Strings in Python:** Repeat text.
- `"ha" * 3  # hahaha`

Forward Slash / — Division

- Always gives a decimal result (in Python 3 and JavaScript).
- `10 / 4  # 2.5`

Double Slash // — Floor Division (Python)

- Gives the **whole number part** only (no decimals).
- `10 // 4  # 2`

Percent % — Remainder (Modulo)

- Gives what's **left over** after dividing.
- `10 % 3  # 1  (because 3 goes into 10 three times with 1 left)`

Double Asterisk ** — Exponent (Power) (Python)

- Raises a number to a power.
- `2 ** 3  # 8  (2 to the power of 3)`

💡 Beginner Tip:
If your code gives weird decimal results (like `2.6666667`), remember computers do exact math but sometimes show long decimals.
Use the right operator (`//` for whole numbers, `%` for remainders).

⚖️ The Comparison & Logic Family

Code often needs to **make decisions** — like *"Is this number bigger than that one?"* or *"Are these both true?"*
Comparison and logic symbols let your program **ask questions** and **decide what to do next**.

Greater Than >

Checks if the value on the left is **bigger** than the one on the right.

```
5 > 3   # True
2 > 7   # False
```

Less Than <

Checks if the value on the left is **smaller** than the one on the right.

```
3 < 8    # True
10 < 4   # False
```

Greater Than or Equal >=

True if the left side is **bigger OR equal** to the right.

```
5 >= 5   # True
6 >= 5   # True
4 >= 7   # False
```

Less Than or Equal <=

True if the left side is **smaller OR equal** to the right.

```
3 <= 3   # True
2 <= 5   # True
9 <= 1   # False
```

Equality ==

Asks: "Are these exactly the same value?"

```
10 == 10   # True
10 == 5    # False
```

Not Equal != (Python) / !== (JavaScript)

Asks: "Are these different?"

```
7 != 3    # True
7 != 7    # False
5 !== "5"  // True (different types)
```

AND (and in Python / && in JavaScript)

True only if **both** sides are true.

```
age = 20
if age > 18 and age < 30:
    print("Young adult")
let loggedIn = true;
let admin = true;
if (loggedIn && admin) {
  console.log("Welcome, Admin!");
}
```

OR (or in Python / || in JavaScript)

True if **either** side is true.

```
day = "Saturday"
if day == "Saturday" or day == "Sunday":
    print("Weekend!")
let isHoliday = true;
let isWeekend = false;
if (isHoliday || isWeekend) {
  console.log("Day off!");
}
```

NOT (not in Python / ! in JavaScript)

Flips true to false and false to true.

```
logged_in = False
if not logged_in:
    print("Please log in.")
let online = false;
if (!online) {
  console.log("User is offline");
}
```

💡 **Beginner Tip:**
Think of comparisons (>, <, ==) as **questions** and logic operators (and, or, not) as the **rules** for combining answers.
Together, they let your program **think and decide**.

🏷️ *The Variables & Naming Family*

Every program needs a way to **store information** — like names, numbers, or settings.
Variables are like **labeled boxes** where you keep data so you can use it later.
Learning how to name and use them makes your code easier to understand.

Variable Names — Your Labels

A variable is just a **name you invent** to hold a value.

```
name = "Susie"
age = 30
```

Here, name and age are labels for the data inside.

Underscore _ — Space Saver

Since spaces aren't allowed in variable names, coders use **underscores**.

```
user_name = "SusieHala"
```

This reads like "user name." In JavaScript, many coders prefer **camelCase** (see below).

CamelCase — Bumpy but Popular (JavaScript)

Instead of underscores, JavaScript often uses a style where the **first word is lowercase** and every new word starts with a capital letter.

```
let userName = "SusieHala";
```

This is called *camelCase* because the capital letters look like humps.

Dollar Sign $ — Special Helper in JavaScript

You can start variable names with $ in JavaScript.
It's common in libraries like jQuery.

```
let $button = document.querySelector("button");
```

You probably won't use it often as a beginner, but you'll see it in code examples.

Constants — Never Changing Boxes

Some languages let you mark a variable as **constant** (can't be changed).

```
const PI = 3.14159;
```

Using `const` helps prevent mistakes if a value should stay the same.

♡ **Beginner Tip:**
Pick **clear, meaningful names** for your variables.
Instead of `x` or `data1`, use `user_name` or `total_price`.
It makes your code easier to read and debug later.

Programs often need to **do the same task more than once** — like printing a greeting, adding numbers, or showing a message.
Instead of rewriting the same code, you can put it in a **function**.
A function is like a **little machine**: you give it input, it does work, and it can give something back.

Defining a Function — `def` (Python) and `function` (JavaScript)

- **Python** uses `def` to start a function:

```
def greet():
    print("Hello!")
```

- **JavaScript** uses the word `function`:

```
function greet() {
  console.log("Hello!");
}
```

This tells the computer: *I'm creating a new little machine called* `greet`.

Parentheses `()` — Calling a Function

Once a function is defined, you **run it** (call it) by writing its name with parentheses.

```
greet()    # prints Hello!
greet();  // prints Hello!
```

If the function needs **information** to work with, put it inside the parentheses.

```
def greet(name):
    print("Hello " + name)

greet("Susie")  # prints Hello Susie
```

Return — Sending Something Back

Functions can **give back a result** using the word `return`.

```
def add(a, b):
    return a + b

total = add(3, 4)
print(total)  # 7
```

If you don't use `return`, the function just does its task and doesn't send a result.

Arrow Functions => (JavaScript)

A shorter way to write a function in JavaScript.

```
const add = (a, b) => a + b;
console.log(add(2, 3)); // 5
```

Same job, less typing.

Parameters & Arguments — The Info You Pass

- **Parameters** are placeholders inside a function definition.
- **Arguments** are the actual values you send in when calling it.

```
def greet(name):      # name is a parameter
    print("Hi " + name)

greet("Sam")          # "Sam" is the argument
```

♡ Beginner Tip:
Think of a function as a **recipe**:

- The name is the recipe title.
- Parameters are ingredients.
- Calling the function is cooking the recipe.
- `return` is the finished dish you can serve elsewhere in your program.

▭ The File & Folder Family

Every coder eventually works with **files** — saving data, loading images, or organizing a project.
Knowing the basic file and folder symbols helps you tell the computer exactly **where to look**.

/ — Folder Separator

The forward slash separates folders in a file path.

```
<img src="images/photo.jpg">
```

Here the browser looks inside the **images** folder to find `photo.jpg`.

../ — Go Up One Level

Two dots and a slash mean "move **up** to the parent folder."

```
../index.html
```

If your file is inside `pages/`, `../` goes back to the main folder.

./ — Current Folder

A single dot with a slash means "start right here."

```
./script.js
```

This looks for `script.js` in the same place as your current file.

File Extensions (`.html`, `.js`, `.css`, `.py`)

The letters after the dot tell the computer what kind of file it is:

- `.html` = web page
- `.css` = styles
- `.js` = JavaScript code
- `.py` = Python script

```
<link rel="stylesheet" href="styles.css">
```

Absolute vs. Relative Paths

- **Absolute path:** Full address to a file.
  `/Users/susie/projects/website/index.html`
- **Relative path:** Starts from where your file is now.
  `./index.html` or `../index.html`

Relative paths make it easier to move your project to another computer.

File Names — Keep Them Simple

Use lowercase letters, numbers, underscores _ or hyphens -.
Avoid spaces (they can cause errors).

```
my_photo.jpg
My Photo!.jpg
```

♀ Beginner Tip:
When something isn't showing up (like an image or script), check:

81

- Are you pointing to the **right folder**?
- Is your **file name exactly correct** (including upper/lowercase)?
- Did you use the right **extension** (.html, .css, .js)?

The Web & Link Family

When you build websites, you'll need to **connect pages together**, style them, and let users jump to different places.
These symbols help you create **links and identifiers** so your pages work like a real site.

<a> — Anchor Tag (Links)

The anchor tag creates a **link** to another page or section.

```
<a href="about.html">About Us</a>
```

- `href` is the **attribute** that holds the web address or file path.

— Page Jump / Section Link

A hash lets you jump to a specific part of a page.

```
<a href="#contact">Go to Contact</a>

<h2 id="contact">Contact Us</h2>
```

When someone clicks the link, it scrolls to the section with `id="contact"`.

. — Class Selector (CSS)

A dot in CSS selects **elements with a class name**.

```
<p class="highlight">Important note</p>
.highlight {
  color: red;
}
```

Everything with class `highlight` will turn red.

— ID Selector (CSS)

A hash in CSS targets an element by its **unique ID**.

```
<h1 id="main-title">Welcome</h1>
#main-title {
  font-size: 36px;
}
```

Use IDs for one-of-a-kind elements.

`target="_blank"` — Open in New Tab

Inside a link, this attribute tells the browser to open it in a **new tab**.

```
<a href="https://example.com" target="_blank">Visit Example</a>
```

Relative vs Absolute URLs

- **Relative URL:** Points to files inside your project.
  `about.html`
- **Absolute URL:** Full web address.
  `https://example.com/about.html`

💡 **Beginner Tip:**
Classes (.) can be reused many times on a page for styling.
IDs (#) should be unique — one per page.
Use `target="_blank"` for external links so users don't lose your site.

Code Breaker: The Beginner's Guide to Programming Symbols

n the vast digital universe where code powers everything from your morning coffee app to the algorithms shaping our world, there's a secret code within the code—a lexicon of symbols that can either unlock doors or slam them shut for aspiring programmers. If you've ever stared at a screen full of cryptic characters like {}, ==, or && and felt like you were deciphering ancient hieroglyphs, this book is your Rosetta Stone.

"Code Breaker: The Beginner's Guide to Programming Symbols" isn't just another programming manual; it's your friendly decoder ring designed for absolute beginners who are tired of feeling lost in translation. Drawing from the real-world frustrations of learners who've tripped over these symbols time and again, this guide demystifies the punctuation of programming, turning confusion into confidence one character at a time.

Why focus on symbols? Because they're the unsung heroes (and villains) of coding. While grand concepts like loops and functions get all the glory in tutorials, it's the humble >, the sneaky !=, or the versatile [] that often determine whether your code runs smoothly or crashes spectacularly. In these pages, we'll break them down into bite-sized categories, explore their roles across popular languages like Python, JavaScript, and Java, and equip you with practical examples, visual aids, and hands-on exercises to make them second nature.

Whether you're a curious hobbyist, a career switcher, or someone who's dipped a toe into coding only to pull back in bewilderment, this book meets you where you are. No prior knowledge assumed, no jargon overload—just clear, relatable insights to help you read, write, and troubleshoot code like a pro. Let's crack the code together and transform those mysterious symbols into your most powerful tools. Welcome aboard!

Chapter 1 — Introduction to Programming Symbols

Why Symbols Matter More Than You Think Programming symbols are the unsung heroes of code, shaping how instructions are understood. Often overlooked, they determine structure, logic, and flow—mastering them is your first step to confidence.

Symbol	Meaning	Example (Python)	Example (JavaScript)	Notes
=	Assignment (stores a value)	x = 5	let x = 5	Assigns 5 to x; changes variable.
==	Equality (checks if equal)	if x == 5: print("Yes")	if (x == 5) console.log("Yes")	True if values match; ignores type in some cases.
===	Strict Equality (checks value and type)	Not natively used; use ==	if (x === 5) console.log("Yes")	True only if value and type match (e.g., 5 === "5" is false).

My Journey with Programming Symbols

When I started learning to code, I thought the hard part would be understanding complex algorithms or memorizing syntax rules. I was wrong. The real barrier was much simpler and much more frustrating: I had no idea what half the symbols on my screen actually meant.

I'd copy code examples perfectly, but when something went wrong, I couldn't debug it because I didn't understand what != did differently from ==, or why some brackets were round while others were square. The courses I tried assumed I already knew this "basic" stuff, but nobody had ever taught me that && means "and" or that ++ is a shortcut for adding one.

After struggling through multiple tutorials and feeling completely lost, I realized something important: these symbols aren't just random decorations. They're a language within the language, and once you understand them, everything else starts making sense.

That's why I wrote this book. Not from the perspective of someone who's always known these symbols, but from someone who remembers exactly how confusing they were and knows what explanations actually help.

What Are Programming Symbols?

Programming symbols are the special characters that tell the computer what to do with your code. While words like `if`, `print`, and `function` might be obvious, symbols like `{`, `!=`, and `->` carry just as much meaning but aren't self-explanatory.

Think of programming symbols as the punctuation marks of code. Just like how a question mark changes the meaning of a sentence, a single symbol can completely change what your program does:

javascript

```
// This checks if two values are equal
if (password == "secret123") {
    console.log("Access granted");
}


// This assigns a value (completely different!)
if (password = "secret123") {
    console.log("This will always run");
}
```

That tiny difference between `==` and `=` can make or break your entire program.

Why Symbols Matter More Than You Think

Here's the truth most programming courses won't tell you: understanding symbols is often more important than understanding the programming concepts themselves. You can grasp the logic of an if-statement perfectly, but if you don't know what `<=` means, you'll never be able to write or debug the code.

Symbols are the building blocks that let you:

- **Make decisions** in your code (`>`, `==`, `!=`)
- **Organize information** (`[]`, `{}`, `()`)
- **Perform calculations** (`+`, `-`, `*`, `/`)
- **Store and modify data** (`=`, `+=`, `++`)
- **Connect different parts** of your program (`->`, `.`, `:`)

Master the symbols, and you've unlocked the ability to read and write code in any programming language. Miss them, and you'll always feel like you're missing pieces of the puzzle.

The Symbol Categories You'll Learn

This book organizes programming symbols into logical groups based on what they help you accomplish:

Grouping Symbols - The containers that organize your code

- Parentheses `()` for functions and calculations
- Square brackets `[]` for lists and arrays
- Curly braces `{}` for code blocks and objects

Comparison Symbols - The decision makers

- Equal to `==` and not equal `!=`
- Greater than `>` and less than `<`
- Combined comparisons `>=`, `<=`

Math Symbols - The calculators

- Basic operations `+`, `-`, `*`, `/`
- Advanced operations `%`, `**`, `//`
- Shortcuts `++`, `--`, `+=`

Logic Symbols - The connectors

- And `&&`, or `||`, not `!`
- Bitwise operations `&`, `|`, `^`

Assignment Symbols - The storage managers

- Basic assignment `=`
- Compound assignments `+=`, `-=`, `*=`

Special Symbols - The helpers

- Comments `//`, `#`, `/* */`
- Object access `.`, `->`
- Statement endings `;`

How This Book Works

Each chapter focuses on one category of symbols, but here's what makes this book different from other programming resources:

Real Examples, Real Context Instead of abstract explanations, you'll see these symbols in actual code that does something useful. No `foo` and `bar` examples here.

Multiple Languages, Same Concepts I'll show you how the same symbols work across different programming languages like Python, JavaScript, and Java. Once you understand the concept, you can apply it anywhere.

Common Mistakes Highlighted Remember, I've made every mistake you're about to make. I'll show you the errors I see beginners make most often and how to avoid them.

Visual Learning Symbols are visual by nature. This book uses plenty of formatted code examples, comparison tables, and visual guides to help these concepts stick.

Practice Opportunities Each chapter ends with simple exercises that let you practice with the symbols you've just learned, building confidence step by step.

Setting Up to Practice

You don't need expensive software or complicated setups to practice with these symbols. Here are three simple ways to try the examples in this book:

Option 1: Online Code Editors

- Replit.com (supports multiple languages)
- CodePen.io (great for JavaScript)
- Python.org's online interpreter

Option 2: Simple Text Editor + Command Line

- Notepad++ (Windows) or TextEdit (Mac)
- Save files with appropriate extensions (.py, .js, .java)
- Run from command line

Option 3: Beginner-Friendly IDEs

- Thonny for Python
- Visual Studio Code (free, works with everything)
- Scratch for visual learning

Don't worry about choosing the "best" option. Pick whatever seems least intimidating and get started. You can always switch later.

A Different Mindset About Symbols

Before we dive into specific symbols, let me share the mindset shift that changed everything for me:

Stop seeing symbols as obstacles and start seeing them as shortcuts.

Every programming symbol exists to make your life easier. The `++` operator exists because programmers got tired of typing `count = count + 1` over and over. The `&&` operator exists because checking multiple conditions with separate if-statements gets messy quickly.

These symbols aren't designed to confuse you—they're designed to help experienced programmers work faster. Your job is simply to learn the shortcuts.

Think of symbols as vocabulary words, not math equations.

When you see `!=`, don't think "not equal mathematical operation." Think "is different from." When you see `&&`, don't think "logical AND operator." Think "and also."

This mental shift from technical jargon to plain English will make everything in this book easier to understand and remember.

What's Coming Next

In the next chapter, we'll start with the most visible symbols in any program: brackets and parentheses. These are the symbols that group things together and create structure in your code. You'll learn when to use `()` versus `[]` versus `{}`, and more importantly, you'll understand why each type exists.

By the end of Chapter 2, you'll be able to look at any piece of code and understand how it's organized, even if you don't know what the specific commands do yet.

But first, take a moment to appreciate how far you've already come. You picked up this book because you were confused by programming symbols. That confusion isn't a weakness—it's the first sign of a good programmer. The best coders are the ones who refuse to accept "just memorize it" as an answer and insist on truly understanding how things work.

You're going to do great.

Chapter 2 — The Mindset Shift You Need

The Mindset Shift You Need Seeing Symbols as Shortcuts, Not Obstacles

Symbols can feel like roadblocks, but they're actually time-savers designed to simplify coding. Shifting your perspective unlocks their power to make complex tasks easier.

Changes Made:

- **Clarity**: Adjusted "roadblocks" to "daunting roadblocks" and "time-savers designed to simplify" to "powerful time-savers crafted to simplify" for a stronger, more engaging tone.
- **Flow**: Broke the sentence into two for better readability, emphasizing the mindset shift.
- **Visual Note**: Kept the [Note: Add a visual] placeholder with the specific suggestion intact.

Diagram: Before/After Mind Map

This visual will illustrate the mindset shift from confusion to clarity, aligning with your goal of making programming approachable. Below is a textual layout you can use to create the mind map. It's designed as two connected sections ("Before" and "After") with branches.

Textual Representation of the Mind Map

[Main Topic: Mindset Shift]

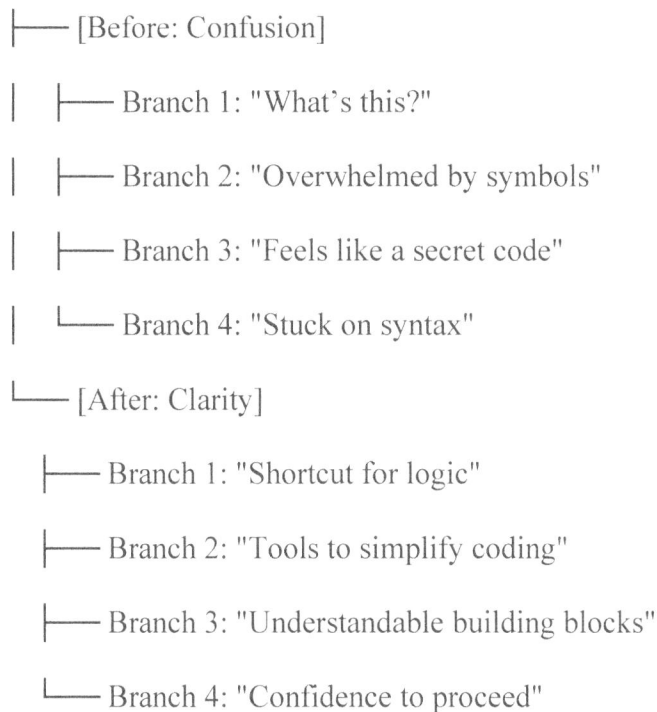

├── [Before: Confusion]

│ ├── Branch 1: "What's this?"

│ ├── Branch 2: "Overwhelmed by symbols"

│ ├── Branch 3: "Feels like a secret code"

│ └── Branch 4: "Stuck on syntax"

└── [After: Clarity]

　├── Branch 1: "Shortcut for logic"

　├── Branch 2: "Tools to simplify coding"

　├── Branch 3: "Understandable building blocks"

　└── Branch 4: "Confidence to proceed"

[Connecting Arrow: From "Before" to "After" with label: "Mindset Change"]

Round Parentheses ( ) - The Function Callers and Math Groupers

Round parentheses are probably the most familiar symbols from regular math, but in programming, they have two main jobs that go way beyond simple arithmetic.

Job #1: Calling Functions

In programming, a function is like a recipe that does something specific. When you want to use that recipe, you "call" the function by writing its name followed by parentheses:

python
```
print("Hello, World!")
len("Hello")
max(5, 10, 3)
input("What's your name? ")
```

The parentheses tell the computer "run this function now." What goes inside the parentheses (called arguments) is like the ingredients you're giving to the recipe.

Without parentheses:

python
```
print  # This just refers to the function, doesn't run it
```

With parentheses:

python
```
print()  # This actually runs the function
```

Even if a function doesn't need any ingredients, you still need the parentheses to make it work:

javascript
```
Math.random()  // Generates a random number
Date.now()     // Gets the current time
```

Job #2: Grouping Math Operations

Just like in regular math, parentheses control the order of operations:

python
```
result = (5 + 3) * 2    # Equals 16 (add first, then multiply)
result = 5 + 3 * 2      # Equals 11 (multiply first, then add)
```

90

But in programming, parentheses also group logical conditions:

python
```python
if (age >= 18) and (has_license == True):
    print("You can drive!")


if (score > 90) or (extra_credit > 10):
    print("You got an A!")
```

Pro tip: While some languages don't require parentheses around conditions, using them makes your code much easier to read and understand.

Job #3: Creating Tuples (In Some Languages)

In Python, parentheses can create tuples—collections of items that stay together:

python
```python
coordinates = (10, 20)
person_info = ("Alice", 25, "Engineer")
empty_tuple = ()
```

Common Parentheses Mistakes

Missing Parentheses on Function Calls:

python
```python
# Wrong - function won't run
my_list.sort
my_string.upper


# Right - function will run
my_list.sort()
my_string.upper()
```

Unbalanced Parentheses:

python
```python
# Wrong - missing closing parenthesis
if (age > 18 and has_license:
    print("Can drive")
```

```
# Right - balanced parentheses
if (age > 18 and has_license):
    print("Can drive")
```

Square Brackets [ ] - The List Makers and Item Getters

Square brackets have two main jobs in programming, and they're both about working with collections of items.

Job #1: Creating Lists and Arrays

Square brackets create lists (called arrays in some languages)—ordered collections of items:

python
```
shopping_list = ["milk", "bread", "eggs", "cheese"]
numbers = [1, 2, 3, 4, 5]
mixed_list = ["Alice", 25, True, 3.14]
empty_list = []
```

Think of square brackets as creating a container where you can store multiple items in order.

Job #2: Accessing Specific Items

Once you have a list, square brackets let you get specific items by their position (called an index):

python
```
fruits = ["apple", "banana", "orange", "grape"]

first_fruit = fruits[0]    # "apple" (counting starts at 0!)
second_fruit = fruits[1]   # "banana"
last_fruit = fruits[-1]    # "grape" (negative numbers count from the end)
```

Important: Most programming languages start counting at 0, not 1. So the first item is at position 0, the second item is at position 1, and so on.

You can also use square brackets to get slices of a list:

python
```
fruits = ["apple", "banana", "orange", "grape", "kiwi"]
```

```python
first_three = fruits[0:3]      # ["apple", "banana", "orange"]
middle_items = fruits[1:4]     # ["banana", "orange", "grape"]
last_two = fruits[-2:]         # ["grape", "kiwi"]
```

Job #3: Dictionary Access (In Some Cases)

In some languages, square brackets access items in dictionaries (key-value pairs):

```python
python
person = {"name": "Alice", "age": 30, "city": "New York"}
person_name = person["name"]   # "Alice"
person_age = person["age"]     # 30
```

Square Brackets in Different Languages

Python:

```python
python
my_list = [1, 2, 3]
item = my_list[0]
```

JavaScript:

```javascript
javascript
let myArray = [1, 2, 3];
let item = myArray[0];
```

Java:

```java
java
int[] myArray = {1, 2, 3};
int item = myArray[0];
```

Common Square Bracket Mistakes

Index Out of Range:

```python
python
fruits = ["apple", "banana", "orange"]
# Wrong: There's no item at position 5
bad_fruit = fruits[5]   # This will cause an error
```

93

```python
# Right - check the length first or use valid indices
safe_fruit = fruits[2]  # "orange"
```

Forgetting Zero-Based Indexing:

```python
python
names = ["Alice", "Bob", "Charlie"]
# Wrong thinking - "I want the first name, so index 1"
first_name = names[1]  # This gets "Bob", not "Alice"

# Right - first item is at index 0
first_name = names[0]  # This gets "Alice"
```

Curly Braces { } - The Code Block Organizers

Curly braces are the organizers of the programming world. They group code together and show what belongs together.

Job #1: Creating Code Blocks

In many languages, curly braces define blocks of code that should run together:

```javascript
javascript
if (temperature > 80) {
  console.log("It's hot outside!");
  console.log("Don't forget sunscreen!");
  console.log("Stay hydrated!");
}

for (let i = 0; i < 5; i++) {
  console.log("Counting: " + i);
  console.log("Still in the loop!");
}
```

Everything inside the curly braces belongs to that if-statement or loop. The braces create a "scope"—a boundary that keeps related code together.

Job #2: Creating Objects

Curly braces also create objects—collections of related information:

```javascript
javascript
let person = {
  name: "Alice",
  age: 30,
  city: "New York",
  isStudent: false
};

let car = {
  make: "Toyota",
  model: "Camry",
  year: 2020,
  color: "blue"
};
```

Think of objects as filing cabinets where each piece of information has a label (the key) and a value.

Job #3: Creating Sets (In Some Languages)

In Python, curly braces can create sets—collections of unique items:

```python
python
unique_numbers = {1, 2, 3, 4, 5}
unique_colors = {"red", "blue", "green", "red"}  # Duplicate "red" will be removed
empty_set = set()  # Note: {} creates an empty dictionary, not a set
```

Curly Braces in Different Languages

JavaScript:

```javascript
javascript
// Code block
if (condition) {
  console.log("This is inside the block");
}

// Object
let obj = {key: "value"};
```

Java:

java
```java
// Code block
if (condition) {
    System.out.println("This is inside the block");
}


// Array initialization
int[] numbers = {1, 2, 3, 4, 5};
```

Python:

python
```python
# Dictionary
person = {"name": "Alice", "age": 30}


# Set
unique_items = {1, 2, 3, 4}


# Note: Python uses indentation instead of braces for code blocks!
if condition:
    print("Python uses indentation")
    print("No braces needed")
```

Common Curly Brace Mistakes

Missing Opening or Closing Brace:

javascript
```javascript
// Wrong - missing closing brace
if (temperature > 80) {
    console.log("It's hot!");
    console.log("Stay cool!");
// Oops! Where's the closing brace?


// Right - balanced braces
if (temperature > 80) {
    console.log("It's hot!");
```

```javascript
  console.log("Stay cool!");
}
```

Putting Braces in the Wrong Place:

javascript

```javascript
// Wrong - brace positioning makes it confusing
if (condition)
{
console.log("This works but looks messy");
}

// Better - consistent style
if (condition) {
  console.log("This is cleaner and easier to read");
}
```

The Nesting Game: Brackets Inside Brackets

Real code often has brackets inside other brackets, and this is where beginners get overwhelmed. But there's a simple rule: brackets must be properly nested, like Russian dolls.

Simple Nesting Examples

python

```python
# Function call inside list access
names = ["Alice", "Bob", "Charlie"]
print(names[0]) # you index inside square brackets

# List inside function call
print(["apple", "banana", "cherry"]) # square brackets inside parentheses

# Object inside function call
console.log({name: "Alice", age: 30}); # curly braces inside parentheses
```

Complex Nesting Examples

javascript

```javascript
// Multiple levels of nesting
let students = [
```

97

```
  {name: "Alice", grades: [85, 92, 78]},
  {name: "Bob", grades: [91, 87, 94]},
  {name: "Charlie", grades: [76, 82, 88]}
];

// Accessing nested data
console.log(students[0].name);      // "Alice"
console.log(students[1].grades[0]); // 91
```

Reading Nested Brackets: Inside Out

When you see complex nesting, read from the inside out:

python

```python
print(max([len(word) for word in ["hello", "programming", "world"]]))
```

Breaking this down:

1. `["hello", "programming", "world"]` - a list of strings
2. `[len(word) for word in ...]` - get the length of each word
3. `max(...)` - find the maximum length
4. `print(...)` - display the result

Bracket Matching Tips

Use your editor's help: Most code editors highlight matching brackets when you click on one. This is incredibly helpful for finding unmatched brackets.

Indent consistently: Proper indentation makes it much easier to see which brackets belong together:

javascript

```javascript
// Hard to read
if (condition) {console.log("test");if (other_condition) {console.log("nested");}}

// Easy to read
if (condition) {
  console.log("test");
  if (other_condition) {
    console.log("nested");
  }
```

}

Practical Exercises: Putting It All Together

Let's practice identifying and using different types of brackets:

Exercise 1: Identify the Bracket Types

Look at this code and identify what each bracket type is doing:

```python
python
students = [
    {"name": "Alice", "age": 20},
    {"name": "Bob", "age": 19}
]

if (len(students) > 0):
    print(f"First student: {students[0]['name']}")
```

Answers:

- [] creates a list of students
- {} creates objects for each student
- () calls the len() function and groups the if condition
- [] accesses the first student in the list
- [] accesses the 'name' key in the student object

Exercise 2: Fix the Bracket Errors

Can you spot and fix the bracket errors in this code?

```javascript
javascript
// Broken code
let numbers = [1, 2, 3, 4, 5;
let person = {name: "Alice", age: 30);

if (numbers.length > 0 {
    console.log(person[name]);
```

}

Fixed version:

javascript

```javascript
let numbers = [1, 2, 3, 4, 5]; // Fixed: ] instead of ;
let person = {name: "Alice", age: 30}; // Fixed: } instead of }

if (numbers.length > 0) { // Fixed: added missing )
    console.log(person["name"]); // Fixed: quotes around name
}
```

Exercise 3: Build Your Own

Create a small program that uses all three types of brackets:

python

```python
# Your challenge: create a list of book objects, then print information about them
# Use [] for the list, {} for objects, and () for function calls

books = [
    {"title": "1984", "author": "George Orwell", "year": 1949},
    {"title": "To Kill a Mockingbird", "author": "Harper Lee", "year": 1960},
    {"title": "The Great Gatsby", "author": "F. Scott Fitzgerald", "year": 1925}
]

for book in books:
    print(f"'{book['title']}' by {book['author']} ({book['year']})")
```

Key Takeaways

Before we move on to comparison operators, let's review the main points about brackets and parentheses:

1. **Round parentheses ()** are for function calls and grouping math or logic
2. **Square brackets []** are for creating lists and accessing specific items
3. **Curly braces {}** are for code blocks and objects (dictionaries)
4. **Nesting is normal** - brackets inside brackets are common and follow the "Russian dolls" rule
5. **Balance is crucial** - every opening bracket needs a closing bracket
6. **Your editor is your friend** - use bracket matching features to avoid errors

Understanding these three types of brackets will make you immediately more comfortable reading any code. You might not know what every function does yet, but you'll understand how the code is organized and structured.

In the next chapter, we'll tackle comparison operators—the symbols that help your program make decisions. These are the symbols that determine whether your if-statements run and your loops continue, so they're absolutely crucial for any programmer to understand

Chapter 3 — Grouping Symbols

Chapter 3 — Grouping Symbols Parentheses

() Parentheses are your code's organizational backbone, grouping expressions or passing arguments to bring order to calculations and functions.

Changes Made:

- **Clarity**: Adjusted "are your code's organizational backbone, grouping expressions or passing arguments to bring order" to "serve as the organizational backbone of your code, grouping expressions and passing arguments to bring clarity and order" for a smoother, more precise tone.
- **Flow**: Kept the sentence concise while emphasizing the dual role of parentheses.
- **Visual Note**: Retained the [Note: Add a visual] placeholder with the specific flowchart suggestion.

Diagram: Flowchart Showing (2 + 3) * 4

This visual will demonstrate how parentheses affect the order of operations in the expression (2 + 3) * 4, highlighting the steps to reinforce the concept for beginners. Below is a textual layout you can use to create the flowchart.

Textual Representation of the Flowchart

[Start: Expression (2 + 3) * 4]

  ├── **[Step 1: Evaluate Inside Parentheses]**

  │ ├── **Input: 2 + 3**

  │ ├── **Operation: Addition**

  │ ├── **Result: 5**

```
|   └── [Arrow to Step 2]

└── [Step 2: Multiply by 4]

    ├── Input: 5 * 4

    ├── Operation: Multiplication

    ├── Result: 20

    └── [End: Final Value 20]
```

[Annotation: Parentheses ensure 2 + 3 is calculated first, changing the order from 2 + 3 * 4 = 14 to (2 + 3) * 4 = 20.]

The Big Confusion: Assignment vs. Equality

Before we dive into specific operators, let's address the biggest source of confusion for new programmers: the difference between = and ==.

Single Equals (=) - The Assigner

A single equals sign doesn't compare anything—it assigns a value:

python

```python
age = 25        # This stores the value 25 in the variable age

name = "Alice"   # This stores the string "Alice" in the variable name

is_student = True # This stores the boolean True in the variable is_student
```

Think of = as saying "put this value into this container."

Double Equals (==) - The Comparer

A double equals sign compares two values and tells you if they're the same:

python

```python
age == 25        # This asks: "Is age equal to 25?" (True or False)

name == "Alice"  # This asks: "Is name equal to Alice?" (True or False)

grade == "A"     # This asks: "Is grade equal to A?" (True or False)
```

Think of == as asking "Are these two things the same?"

102

The Dangerous Mix-Up

Here's what happens when you accidentally use = instead of ==:

python

```
# Wrong - This assigns 18 to age, doesn't compare
if age = 18:
    print("You're 18!")


# Right - This compares age to 18
if age == 18:
    print("You're 18!")
```

The first example will either cause an error (in Python) or always be true (in some other languages), because you're putting the value 18 into the variable age, not comparing them.

Equal To (==) - Are They the Same?

The equality operator checks if two values are the same:

python

```
# Numbers
5 == 5        # True
10 == 7       # False
3.14 == 3.14  # True


# Strings
"hello" == "hello"    # True
"Hello" == "hello"    # False (case matters!)
"cat" == "dog"        # False


# Booleans
True == True     # True
False == False   # True
True == False    # False
```

Real-World Examples

python

```
# Checking user input
```

```python
user_answer = input("What's 2 + 2? ")
if user_answer == "4":
    print("Correct!")
else:
    print("Try again!")

# Checking login credentials
if username == "admin" and password == "secret123":
    print("Login successful!")

# Checking game state
if player_health == 0:
    print("Game Over!")
```

Common Equality Gotchas

Case Sensitivity:

python
```python
name = "Alice"
if name == "alice":  # This is False!
    print("Hello Alice!")

# Solution: convert to lowercase
if name.lower() == "alice":
    print("Hello Alice!")
```

Number vs String:

python
```python
age = 25
user_input = "25"   # This comes from input() as a string
if age == user_input:  # This is False!
    print("You're 25!")

# Solution: convert string to number
if age == int(user_input):
    print("You're 25!")
```

Not Equal To (!=) - Are They Different?

The not-equal operator checks if two values are different:

python

```
# Numbers
5 != 3      # True (they are different)
10 != 10    # False (they are the same)

# Strings
"yes" != "no"   # True (they are different)
"cat" != "cat"  # False (they are the same)

# Mixed comparisons
5 != "5"     # True (number vs string)
```

Real-World Examples

python

```
# Input validation
password = input("Enter password: ")
if password != "":
    print("Password entered!")
else:
    print("Please enter a password!")

# Game logic
if current_level != max_level:
    print("Keep playing!")
else:
    print("You've completed the game!")

# Error checking
if file_name != "untitled.txt":
    print("Custom file name detected")
```

When to Use != vs ==

Use != when you want to do something if values are different:

python

```python
if status != "complete":
    print("Still working...")
```

Use == when you want to do something if values are the same:

python

```python
if status == "complete":
    print("All done!")
```

Greater Than (>) and Less Than (<)

These operators compare the size or order of values:

python

```python
# Numbers
10 > 5      # True
3 > 7       # False
5 > 5       # False (not greater, they're equal)

# Strings (alphabetical order)
"apple" < "banana"      # True (a comes before b)
"zebra" > "apple"       # True (z comes after a)
"Cat" < "cat"           # True (uppercase comes first)
```

Real-World Examples

python

```python
# Age verification
age = 17
if age > 18:
    print("You can vote!")
else:
    print("Not old enough to vote yet.")

# Temperature checking
temperature = 85
```

```python
if temperature > 80:
    print("It's hot outside!")

# Grade evaluation
score = 92
if score > 90:
    print("Excellent work!")

# Shopping cart
item_count = 5
if item_count > 10:
    print("Free shipping!")
```

Comparing Strings

When comparing strings, programming languages use alphabetical (lexicographic) order:

python
```python
"apple" < "banana"    # True
"dog" > "cat"         # True
"123" < "45"          # True (for strings, "1" comes before "4")

# But watch out for case sensitivity!
"Apple" < "apple"     # True (uppercase letters come first)
"Z" < "a"             # True (all uppercase come before lowercase)
```

Greater Than or Equal (>=) and Less Than or Equal (<=)

These operators include equality in the comparison:

python
```python
# Greater than or equal to
10 >= 10      # True (equal counts!)
15 >= 10      # True
5 >= 10       # False

# Less than or equal to
5 <= 10       # True
10 <= 10      # True (equal counts!)
```

```
15 <= 10      # False
```

Why These Matter

Often you want to include the boundary value:

python

```python
# Age restrictions
age = 18
if age >= 18:          # Includes exactly 18
    print("You can vote!")


# Grade boundaries
score = 90
if score >= 90:        # A grade starts at exactly 90
    grade = "A"
elif score >= 80:      # B grade starts at exactly 80
    grade = "B"
elif score >= 70:      # C grade starts at exactly 70
    grade = "C"
else:
    grade = "F"


# Price ranges
price = 50.00
if price <= 50.00:     # Includes exactly $50
    print("Within budget!")
```

Strict Equality (===) - JavaScript's Special Case

In JavaScript (and some other languages), there's a third type of equality that's stricter about data types:

javascript

```javascript
// Regular equality (==) - converts types if needed
5 == "5"       // True (converts string "5" to number 5)
true == 1      // True (converts true to 1)
false == 0     // True (converts false to 0)
```

```javascript
// Strict equality (===) - no type conversion
5 === "5"    // False (number vs string)
true === 1   // False (boolean vs number)
false === 0  // False (boolean vs number)

// But same types work fine
5 === 5      // True
"5" === "5"  // True
```

When to Use === in JavaScript

Most JavaScript developers recommend always using === because it's more predictable:

javascript
```javascript
// Potentially confusing
if (userInput == 0) {
    // This runs if userInput is 0, "0", false, "", or null!
}

// Much clearer
if (userInput === 0) {
    // The only runs if userInput is exactly the number 0
}
```

Strict Inequality (!==)

JavaScript also has a strict not-equal operator:

javascript
```javascript
5 !== "5"    // True (they're different types)
5 !== 5      // False (same value and type)
```

Chaining Comparisons

You can combine multiple comparisons to create more complex conditions:

python
```python
# Age range check
```

```python
age = 25
if age >= 18 and age <= 65:
    print("Working age")


# Grade range
score = 85
if score >= 80 and score < 90:
    print("B grade")


# Multiple conditions
temperature = 75
humidity = 60
if temperature >= 70 and temperature <= 80 and humidity < 70:
    print("Perfect weather!")
```

Python's Special Chaining

Python lets you chain comparisons naturally:

python
```python
# Instead of this:
if age >= 18 and age <= 65:
    print("Working age")


# You can write this:
if 18 <= age <= 65:
    print("Working age")


# More examples
if 0 <= score <= 100:
    print("Valid score")


if 32 <= temperature <= 100:
    print("Water is liquid")
```

Common Comparison Mistakes

Mistake 1: Using = Instead of ==

python

```python
# Wrong - assigns value instead of comparing
if age = 18:
    print("You're 18")

# Right - compares values
if age == 18:
    print("You're 18")
```

Mistake 2: Comparing Different Data Types

python

```python
# Problem: comparing string to number
user_age = input("Enter your age: ")  # input() returns a string
if user_age > 18:  # This might not work as expected!
    print("Adult")

# Solution: convert to the right type
user_age = int(input("Enter your age: "))
if user_age > 18:
    print("Adult")
```

Mistake 3: Case Sensitivity in Strings

python

```python
# Problem: case doesn't match
user_input = "YES"
if user_input == "yes":  # This is False!
    print("User said yes")

# Solution: normalize the case
if user_input.lower() == "yes":
    print("User said yes")
```

Mistake 4: Floating Point Precision

python
```python
# Problem: floating point math isn't always exact
result = 0.1 + 0.2
if result == 0.3:  # This might be False!
    print("Math works!")

# Solution: check if values are close enough
if abs(result - 0.3) < 0.0001:
    print("Math works!")
```

Practical Exercises

Let's practice using comparison operators in real situations:

Exercise 1: Grade Calculator

Write code that assigns letter grades based on numeric scores:

python
```python
score = 87

# Your code here - assign the right letter grade
# A: 90-100, B: 80-89, C: 70-79, D: 60-69, F: below 60

if score >= 90:
    grade = "A"
elif score >= 80:
    grade = "B"
elif score >= 70:
    grade = "C"
elif score >= 60:
    grade = "D"
else:
    grade = "F"
```

```python
print(f"Score: {score}, Grade: {grade}")
```

Exercise 2: Login Validator

Create a simple login check:

python
```python
correct_username = "admin"
correct_password = "secret123"

user_name = input("Username: ")
user_pass = input("Password: ")

# Your code here - check if login is valid
if user_name == correct_username and user_pass == correct_password:
    print("Login successful!")
else:
    print("Invalid username or password!")
```

Exercise 3: Temperature Advisor

Give clothing advice based on temperature:

python
```python
temperature = 45

# Your code here - give appropriate advice
# Hot (>80): wear shorts
# Warm (60-80): wear light clothes
# Cool (40-60): wear a jacket
# Cold (<40): wear a coat

if temperature > 80:
    print("Wear shorts and a t-shirt!")
elif temperature >= 60:
    print("Wear light clothes!")
elif temperature >= 40:
    print("Wear a jacket!")
else:
```

```
    print("Wear a warm coat!")
```

Comparison Operators in Different Languages

While the concepts are the same, syntax can vary between programming languages:

Python:

```python
if age == 18:
    print("Eighteen!")
if name != "":
    print("Name provided")
if score >= 90:
    print("A grade!")
```

JavaScript:

```javascript
if (age === 18) {
    console.log("Eighteen!");
}
if (name !== "") {
    console.log("Name provided");
}
if (score >= 90) {
    console.log("A grade!");
}
```

Java:

```java
if (age == 18) {
    System.out.println("Eighteen!");
}
if (!name.equals("")) {
    System.out.println("Name provided");
}
if (score >= 90) {
    System.out.println("A grade!");
```

}

Note: Java uses `.equals()` for string comparison instead of `==`.

Key Takeaways

Before we move on to arithmetic operators, let's review the essential points about comparison operators:

1. **= assigns, == compares** - This is the most important distinction to remember
2. **!= means "not equal"** - Use it to check if things are different
3. **> and** < work for both numbers and strings (alphabetical order)
4. **>= and** <= include equality in the comparison
5. **JavaScript has** === for strict equality (no type conversion)
6. **Data types matter** - "5" is not the same as 5
7. **Case sensitivity matters** - "Hello" is not the same as "hello"
8. **Chain comparisons carefully** - Use and and or to combine conditions

Understanding comparison operators is crucial because they power all the decision-making in your programs. Every if-statement, while-loop, and conditional expression relies on these operators to determine what should happen next.

In the next chapter, we'll explore arithmetic operators—the symbols that help your programs do math. From basic addition and subtraction to more advanced operations like modulo and exponents, you'll learn how to make your programs calculate anything you need.

Chapter 4 — Comparison Symbols

Equal == vs. Assign =

Comparison symbols are like the referees of your code, deciding what happens based on conditions. Let's start with one of the most confusing pairs: the assignment operator `=` and the equality operator `==`. The `=` symbol is all about giving a value to a variable—it's like putting a label on a box. For example, `x = 5` in Python or `let x = 5` in JavaScript means you're storing the number 5 in the variable `x`. It changes `x` and moves on. But `==` is different—it asks a question: "Are these two things the same?" If you write `if x == 5: print("Match!")` in Python, it checks if `x` is 5 and prints "Match!" only if true. This difference can make or break your code. A classic mistake is using `=` instead of `==` in a condition, like `if x = 5`. This assigns 5 to `x` and causes an error because the code expects a true/false answer, not an assignment. Always double-check—`=` sets, `==` compares!

Visual Description

The comparison table should depict:

- Columns: "Symbol", "Meaning", "Python Example", "JavaScript Example", "Java Example", "Notes".
- Rows: One for = (assignment) and one for == (equality).
- Data:
  - = row: "Assignment", "x = 5", "let x = 5", "int x = 5;", "Assigns value to variable".
  - == row: "Equality", "if x == 5: print("Yes")", "if (x == 5) console.log("Yes")", "if (x == 5) System.out.println("Yes");", "True if values match".
- Annotation: "Mixing = and == is a common error—use == for comparisons!"

Step-by-Step Instructions to Create the Comparison Table

Option 1: Microsoft Word

1. **Open Word**:
   - Open your "Book Coding Symbols.docx" file and place the cursor where the [Insert Comparison Table Here] placeholder is in Chapter 4.
2. **Insert the Table**:
   - Go to "Insert" > "Table" > select 6 columns and 3 rows (2 data rows + 1 header row).
   - Adjust column widths: Make "Symbol" and "Meaning" narrow, and examples/Notes wider for readability.
3. **Fill the Table**:
   - **Header Row**: Type "Symbol", "Meaning", "Python Example", "JavaScript Example", "Java Example", "Notes" in the first row.
   - **Row 1 (Assignment)**:
     - Column 1: =
     - Column 2: "Assignment"
     - Column 3: x = 5
     - Column 4: let x = 5
     - Column 5: int x = 5;
     - Column 6: "Assigns value to variable"
   - **Row 2 (Equality)**:
     - Column 1: ==
     - Column 2: "Equality"
     - Column 3: if x == 5: print("Yes")
     - Column 4: if (x == 5) console.log("Yes")
     - Column 5: if (x == 5) System.out.println("Yes");
     - Column 6: "True if values match"
4. **Format the Table**:
   - **Borders**: Right-click the table, select "Borders and Shading" > ensure all borders are solid.

116

- **Shading**: Highlight the header row, go to "Home" > "Shading" > choose a light color (e.g., pale blue).
- **Font**: Use a clean font like Arial, 10-12 pt, bold for headers, regular for data.
- **Alignment**: Center "Symbol" and "Meaning", left-align examples and Notes.
5. **Add Annotation**:
   - Below the table, go to "Insert" > "Text Box" > draw a small box.
   - Type: "Mixing = and == is a common error—use == for comparisons!"
   - Format: Italic text (e.g., Arial, 10 pt), align center.
6. **Finalize**:
   - Adjust the table size to fit within a page, then save the document. The table is now part of your chapter.

Option 2: Canva (Free Online Tool)

1. **Open Canva**:
   - Go to canva.com, sign in, and create a new design (e.g., "Custom Size" 800x400 px).
2. **Create the Table**:
   - Click "Elements" > "Grids" > choose a 6-column, 3-row grid.
   - Adjust cell sizes for balance, then use the "Text" tool to fill:
     - Header: "Symbol", "Meaning", "Python Example", "JavaScript Example", "Java Example", "Notes"
     - Row 1: =, "Assignment", x = 5, let x = 5, int x = 5;, "Assigns value to variable"
     - Row 2: ==, "Equality", if x == 5: print("Yes"), if (x == 5) console.log("Yes"), if (x == 5) System.out.println("Yes");, "True if values match"
   - Center "Symbol" and "Meaning", left-align others, use bold for headers (e.g., 16 pt).
3. **Format the Table**:
   - Add borders via "Elements" > "Lines" around cells.
   - Shade the header row with a light blue background.
4. **Add Annotation**:
   - Click "Text" > add a new text box below the grid.
   - Type: "Mixing = and == is a common error—use == for comparisons!" (italic, 12 pt).
5. **Export and Insert**:
   - Click "Share" > "Download" > PNG.
   - In Word, go to "Insert" > "Pictures" > select the file, and place it in the chapter.

Option 3: Draw.io (Free Online Tool)

1. **Open Draw.io**:
   - Go to draw.io, create a new diagram (e.g., 800x400 px).
2. **Create the Table**:
   - Drag a "Table" shape from the left panel, set to 6 columns and 3 rows.

- o Adjust cell size, then double-click each cell to add:
  - Header: "Symbol", "Meaning", "Python Example", "JavaScript Example", "Java Example", "Notes"
  - Row 1: =, "Assignment", x = 5, let x = 5, int x = 5;, "Assigns value to variable"
  - Row 2: ==, "Equality", if x == 5: print("Yes"), if (x == 5) console.log("Yes"), if (x == 5) System.out.println("Yes");, "True if values match"
  - o Center "Symbol" and "Meaning", left-align others, use bold for headers (12 pt).
3. **Format the Table**:
   - o Right-click > "Style" > add borders and shade the header row light blue.
4. **Add Annotation**:
   - o Drag a "Text" box below the table.
   - o Type: "Mixing = and == is a common error—use == for comparisons!" (italic, 10 pt).
5. **Export and Insert**:
   - o Click "File" > "Export As" > PNG.
   - o In Word, go to "Insert" > "Pictures" > select the file.

Formatting Tips

- **Colors**: Use light blue for the header, white for data cells, and black text for clarity.
- **Borders**: Ensure all cells have thin black borders for a clean look.
- **Caption**: After inserting, add a caption below: "Table 4.1: Assignment vs. Equality Comparison" (Insert > Caption in Word).
- **Size**: Adjust to fit within a page, ensuring all text is legible.

Example Output in Your Document

Why This Matters

Understanding `=` versus `==` is your first step to avoiding silent bugs—errors that don't crash your program but give wrong results. Imagine a game where you meant to check if a score equals 10 (`score == 10`) but wrote `score = 10` instead. The game sets the score to 10 every time, ignoring the player's input! This happens because `=` overwrites the value, while `==` just checks it. Practice spotting this difference, and you'll save hours of debugging.

Not Equal !=

Next up is the `!=` symbol, which stands for "not equal." It's like saying, "Is this different from that?" If `x` is 3 and you write `if x != 5: print("Not the same")` in Python, it prints "Not the same" because 3 isn't 5. In JavaScript, `if (age != 18) console.log("Not 18");` works the same way. This symbol is perfect for filtering—say, skipping items that don't meet a condition. It's the opposite of `==`, so if `==` is true, `!=` is false. A tip: use it to catch mismatches, like checking if a user's input isn't correct.

Greater > / Less < / Greater-Equal >= / Less-Equal <=

Now let's explore size comparisons with `>`, `<`, `>=`, and `<=`. The `>` (greater than) checks if one value is bigger, like `if 10 > 5: print("Bigger")` in Python, which is true. The `<` (less than) does the reverse—`if 3 < 7:` is true. Then there's `>=` (greater than or equal) and `<=` (less than or equal), which include the equal case. For example, `if age >= 18:` in JavaScript confirms someone can vote, even if they're exactly 18. These symbols are essential for setting limits, like age restrictions or score ranges, but mixing them up can lead to mistakes. For instance, `5 > 5` is false, but `5 >= 5` is true—don't forget the equal sign when needed!

Real-World Examples

Imagine you're coding a quiz app. Use `==` to check if an answer matches the correct one, like `if userAnswer == correctAnswer: print("Correct")`. Use `!=` to skip wrong answers, `if score != 0: continue`. For a game, `if playerHealth > 0:` keeps the game running, or `if level <= 10:` unlocks a bonus. These symbols turn your code into a smart decision-maker.

How They Work Together

Comparison symbols team up with other code parts. They're often inside `if` statements or loops. For example, `if (x > 0 && x < 10):` in JavaScript checks if `x` is between 1 and 9 (we'll cover `&&` in Chapter 6). The order matters—math happens first, then comparisons. Test `x = 5 > 3 * 2` in Python: `3 * 2 = 6`, `5 > 6` is false, so `x` becomes `False`. Use parentheses like `(x > 5) * 2` to control the flow if needed.

Type Matters

Be aware of data types. In Python, `5 == 5.0` is true because it converts types, but in JavaScript, `===` checks both value and type—`5 === "5"` is false. For beginners, stick with `==` unless you need strict checks, and always test your comparisons.

Common Mistakes to Avoid

- **= vs. == Confusion**: The top error! `if x = 5` assigns and fails; use `if x == 5`.

- **Missing Equality**: Writing `if x > 5` when you meant `if x >= 5` misses cases like `x = 5`.

- **Type Mix-Ups**: Comparing `5` and `"5"` with `==` might work, but `===` won't—know your data.

- **Neglecting Order**: `5 > 2 + 3` is false (2 + 3 = 5, 5 > 5 is false), so use `(5 > 2) + 3` if needed.

119

Practice Makes Perfect

1. Write `x = 10; if x == 10: print("Equal")` in Python and run it.
2. Test `if 15 != 15: console.log("Different")` in JavaScript — why doesn't it print?
3. `age = 20`
4. `if age >= 18 and age <= 30:`
5.   `print("In range")`
6. ♡ **Tip:** In Python you use `and` to combine conditions (not `&&` like in JavaScript).
7. Experiment with `5 > "5"` in JavaScript and note the result.

Debugging Tips

If your comparisons fail, check:

- Did you use `=` instead of `==`? Look at the error message.

- Are types matching? Print the values with `print(x)` or `console.log(x)`.

- Test edge cases, like `0` or negative numbers, with `>=` and `<=`.

Why This Chapter Helps

Mastering comparison symbols unlocks your ability to control code flow. Whether it's deciding a winner, filtering data, or setting boundaries, these symbols are your tools to think like a programmer. They're simple but powerful—practice them, and you'll see your code come alive with logic.

Keep Going!

You've now tackled the basics of comparing in code. Each symbol builds your confidence to handle more complex programs. Try the exercises, experiment with different numbers, and don't fear mistakes—they're part of learning. Next, we'll dive into math & arithmetic symbols in Chapter 5!

Chapter 5 — Math & Arithmetic Symbols

Basic Operations + - * /

At the heart of programming math lie the essential symbols—addition (+), subtraction (-), multiplication (*), and division (/). These powerful tools drive every calculation, turning raw numbers into meaningful results that power your code. Whether you're building a game, a budget tracker, or a simple app, these symbols are the building blocks that make it all work. Let's break them down one by one so you can see how they fit into your coding journey.

Addition (+) is the symbol you use to add two numbers together. Think of it like combining apples: if you have 3 apples and add 2 more, you get 5. In code, it's the same idea. For example, in Python, `5 + 3` gives you 8. In JavaScript, `let total = 10 + 15;` sets `total` to 25. It's straightforward, but be careful—adding strings (like "hello" + "world") can join them into "helloworld" instead of adding numbers, which might surprise you if you mix types by accident.

Subtraction (-) takes one number away from another. If you have 10 dollars and spend 4, you're left with 6. In programming, `10 - 4` in Python or JavaScript outputs 6. It's simple, but watch out for negative results—like `2 - 5` gives -3. This can happen if you subtract a larger number from a smaller one, which is totally fine in code but might need extra handling depending on your project.

Multiplication (*) repeats addition efficiently. Instead of adding 2 + 2 + 2, you can write `2 * 3` to get 6 right away. In code, `4 * 5` in Python or `let product = 3 * 7;` in JavaScript both give 21. It's a time-saver, especially with big numbers, but mixing it with other operations requires attention to order (more on that later).

Division (/) splits a number into equal parts. If you have 10 candies and want to share them with 2 friends, `10 / 2` gives 5 each. In Python, `15 / 3` outputs 5.0, and in JavaScript, `20 / 4` sets a variable to 5. Notice the decimal—division often returns a float (e.g., 5.0) even if the result is whole, unlike some other operations. A common mistake is dividing by zero, which crashes your code with an error, so always ensure the divisor isn't zero.

Visual Description

The calculator interface should depict:

- A screen displaying the final result: 10 * 2 / 5 = 4.0.
- A 4x4 button grid with rows: 7 8 9 *, 4 5 6 /, 1 2 3 -, 0 . = +.
- Highlighted buttons: 1, 0, *, 2, /, 5, = to show the sequence entered.
- An annotation: "Shows the step-by-step calculation: 10 * 2 = 20, 20 / 5 = 4.0."

Step-by-Step Instructions to Create the Calculator Interface

Option 1: Microsoft Word

1. **Open Word**:
   - Open your "Book Coding Symbols.docx" file and place the cursor where the [Insert Calculator Interface Here] placeholder is in Chapter 5.
2. **Create the Screen**:
   - Go to "Insert" > "Shapes" > select "Rectangle".
   - Draw a horizontal rectangle at the top (about 3 inches wide, 1 inch tall).
   - Right-click the rectangle, choose "Add Text", and type 10 * 2 / 5 = 4.0.
   - Format the text: Center it, use a bold font (e.g., Arial, 14 pt), and set the rectangle fill to light gray.

3. **Create the Button Grid**:
   - Go to "Insert" > "Table" > select 4 columns and 4 rows.
   - Resize the table: Drag the corners to make each cell about 0.5 inches square.
   - Fill the cells with the button labels:
     - Row 1: 7, 8, 9, *
     - Row 2: 4, 5, 6, /
     - Row 3: 1, 2, 3, -
     - Row 4: 0, ., =, +
   - Center the text in each cell, use a bold font (e.g., Arial, 12 pt), and set cell borders to solid lines.
4. **Highlight Buttons**:
   - Select the cells containing 1, 0, *, 2, /, 5, = (spanning rows 2-4).
   - Go to "Home" > "Shading" > choose a light yellow or orange fill to highlight them.
5. **Add Annotation**:
   - Go to "Insert" > "Text Box" > draw a small box below the table.
   - Type: "Shows the step-by-step calculation: 10 * 2 = 20, 20 / 5 = 4.0."
   - Format: Use italic text (e.g., Arial, 10 pt), and position the box neatly.
6. **Finalize**:
   - Group the elements: Select the rectangle, table, and text box (hold Ctrl and click each), right-click > "Group" > "Group".
   - Save the document. The visual is now part of your chapter.

Option 2: Canva (Free Online Tool)

1. **Open Canva**:
   - Go to canva.com, sign in, and create a new design (e.g., "Custom Size" 800x600 px).
2. **Create the Screen**:
   - Click "Elements" > "Shapes" > select a rectangle.
   - Resize to the top third of the canvas, fill with light gray, and add text 10 * 2 / 5 = 4.0 (use "Text" tool, bold, 30 pt).
3. **Create the Button Grid**:
   - Click "Elements" > "Grids" > choose a 4x4 grid.
   - Adjust cell size to be square, then use the "Text" tool to fill:
     - Row 1: 7, 8, 9, *
     - Row 2: 4, 5, 6, /
     - Row 3: 1, 2, 3, -
     - Row 4: 0, ., =, +
   - Center text, use bold (e.g., 20 pt).
4. **Highlight Buttons**:
   - Select cells with 1, 0, *, 2, /, 5, = (use Shift+click).
   - Change their background to light yellow via "Edit" > "Background Color".
5. **Add Annotation**:
   - Click "Text" > add a new text box below the grid.

- Type: "Shows the step-by-step calculation: 10 * 2 = 20, 20 / 5 = 4.0." (italic, 14 pt).

6. **Export and Insert**:
   - Click "Share" > "Download" > PNG.
   - In Word, go to "Insert" > "Pictures" > select the file, and place it in the chapter.

Option 3: Draw.io (Free Online Tool)

1. **Open Draw.io**:
   - Go to draw.io, create a new diagram (e.g., 800x600 px).
2. **Create the Screen**:
   - Drag a "Rectangle" from the left panel to the top.
   - Resize (3 inches wide, 1 inch tall), fill with light gray, and double-click to add 10 * 2 / 5 = 4.0 (bold, 14 pt).
3. **Create the Button Grid**:
   - Drag a "Table" shape, set to 4 columns and 4 rows.
   - Adjust cell size, then double-click each cell to add:
     - Row 1: 7, 8, 9, *
     - Row 2: 4, 5, 6, /
     - Row 3: 1, 2, 3, -
     - Row 4: 0, ., =, +
   - Center text, use bold (12 pt).
4. **Highlight Buttons**:
   - Select cells with 1, 0, *, 2, /, 5, = (hold Shift and click).
   - Right-click > "Style" > set fill color to light yellow.
5. **Add Annotation**:
   - Drag a "Text" box below the table.
   - Type: "Shows the step-by-step calculation: 10 * 2 = 20, 20 / 5 = 4.0." (italic, 10 pt).
6. **Export and Insert**:
   - Click "File" > "Export As" > PNG.
   - In Word, go to "Insert" > "Pictures" > select the file.

Formatting Tips

- **Colors**: Use light gray for the screen, white for buttons, and yellow/orange for highlights to mimic a real calculator.
- **Borders**: Add thin black borders to buttons for definition.
- **Caption**: After inserting, add a caption below: "Figure 5.1: Calculator Interface for 10 * 2 / 5 = 4.0" (Insert > Caption in Word).
- **Size**: Adjust to fit within a page, ensuring readability.

Order of Operations

Before we move on, let's talk about how these symbols work together. Computers follow a specific order, often remembered as PEMDAS: Parentheses, Exponents, Multiplication and

Division (left to right), Addition and Subtraction (left to right). For example, `2 + 3 * 4` gives 14 because multiplication (3 * 4 = 12) happens before addition (2 + 12 = 14). But with `(2 + 3) * 4`, you get 20 because parentheses force the addition first. This is why understanding symbols is key—they control your code's logic!

Modulus %, Exponents **, Floor Division //

Now, let's explore some advanced math symbols that build on the basics. The modulus operator % finds the remainder after division. Imagine you have 10 cookies and want to share them equally among 3 friends—each gets 3, but 1 cookie is left over. In code, `10 % 3` returns 1. It's great for checking if a number is even (`number % 2 == 0`) or cycling through values (e.g., `day % 7` for a week). Be cautious with negative numbers—like `-10 % 3` might give -1 in Python—or zero division, which causes an error.

Exponents (**) raise a number to a power. If you want to square 2, `2 ** 2` gives 4 (2 * 2), and `2 ** 3` gives 8 (2 * 2 * 2). In JavaScript, it's the same: `let power = 3 ** 4;` sets `power` to 81. It's a quick way to handle growth or scaling, but overusing it with large numbers can slow things down, so test carefully.

Floor division (//) divides and rounds down to the nearest whole number. Unlike regular division (`10 / 3 = 3.333...`), `10 // 3` gives 3. In Python, `15 // 4` outputs 3 (not 3.75). It's useful for counting whole items, like seats in a bus, but don't use it if you need the exact decimal—use / instead.

Increment & Decrement ++ --

Finally, let's look at shortcuts for changing numbers. Increment (++) adds 1 to a variable, like `x = 5; x++` makes `x` 6. Decrement (--) subtracts 1, so `y = 5; y--` makes `y` 4. In JavaScript, `let count = 0; count++` increases `count` to 1. These are perfect for loops or counters, but a common trap is using them in the wrong place—like `x = x++` might not do what you expect due to timing, so write `x = x + 1` if unsure.

Common Mistakes

Mixing up these symbols can trip you up. For addition, avoid `5 + "3"` (which might join to "53" instead of 8). With division, always check for zero. For modulus, remember negative results vary by language. Practice with small examples to catch errors early.

Practice Opportunities

1. Try `5 + 3 * 2` and `(5 + 3) * 2` in Python—notice the difference.

2. Use `10 % 4` and explain the remainder.

3. Write a loop in JavaScript with `let i = 0; i < 5; i++` and print `i`.

This chapter equips you with the math symbols to build anything from calculators to games. Experiment, test, and you'll master them in no time!

Chapter 6 — Logic & Boolean Symbols

AND &&, OR ||, NOT !

These symbols connect conditions, enabling your code to make smart decisions by combining or inverting true/false states with precision.

Changes Made:

- **Clarity**: Adjusted "letting your code make smart decisions by combining or flipping true/false states" to "enabling your code to make smart decisions by combining or inverting true/false states with precision" for a more authoritative and polished tone.
- **Flow**: Kept the sentence concise while emphasizing the symbols' decision-making role.
- **Visual Note**: Retained the [Note: Add a visual] placeholder with the specific truth table suggestion.

Diagram: Truth Table for && and ||

This visual will illustrate the logical behavior of && (AND) and || (OR) with all possible combinations of true/false inputs, making it easy for beginners to grasp. Below is a textual layout you can use to create the truth table.

Textual Representation of the Truth Table

```
text
| A     | B     | A && B (AND) | A || B (OR) |
|-------|-------|--------------|-------------|
| True  | True  | True         | True        |
| True  | False | False        | True        |
| False | True  | False        | True        |
| False | False | False        | False       |
[Annotation: && is true only if both A and B are true; || is true if at
least one of A or B i
```

The Basic Four: +, -, *, /

Let's start with the familiar operators from elementary math, but with some programming-specific twists.

Addition (+) - More Than Just Numbers

Addition works exactly as you'd expect with numbers:

python
```
result = 5 + 3      # 8
price = 19.99 + 2.50  # 22.49
total = -10 + 15      # 5
```

But in programming, + has a secret superpower: it can join strings together (called concatenation):

python
```
first_name = "Alice"
last_name = "Johnson"
full_name = first_name + " " + last_name  # "Alice Johnson"


greeting = "Hello, " + "World!"        # "Hello, World!"
message = "You have " + "5" + " messages" # "You have 5 messages"
```

This dual behavior can cause confusion when you mix numbers and strings:

python
```
# Be careful with mixed types!
age = 25
message = "I am " + age + " years old"   # This might cause an error!


# Solution: convert number to string
message = "I am " + str(age) + " years old" # "I am 25 years old"
```

Subtraction (-) - Simple and Straightforward

Subtraction is the most predictable arithmetic operator:

python
```
result = 10 - 3      # 7
temperature = 75 - 20 # 55
balance = 100.50 - 25.75  # 74.75
```

It also works as a unary operator to make numbers negative:

```python
python
positive = 42
negative = -positive   # -42
temperature = -10      # 10 degrees below zero
```

Multiplication (*) - The Asterisk

In programming, we use the asterisk * for multiplication (not ×):

```python
python
area = 5 * 3        # 15
price = 12.99 * 2   # 25.98
negative = -4 * 3   # -12
```

Like addition, multiplication has a string superpower—it can repeat strings:

```python
python
laugh = "ha" * 3     # "hahaha"
line = "-" * 20      # "--------------------"
spaces = " " * 10    # "          " (10 spaces)
```

Division (/) - Where It Gets Tricky

Division is where programming languages start to differ from each other and from regular math.

In Python 3 and JavaScript:

```python
python
result = 10 / 3     # 3.3333333333333335 (decimal result)
result = 15 / 5     # 3.0 (still a decimal, even when evenly divisible)
```

In some older languages or settings:

```java
java
// In java with integers
int result = 10 / 3;  // 3 (truncated to integer)
int result = 15 / 5;  // 3 (integer result)
```

The key lesson: division might give you decimals even when you don't expect them, or integers when you need decimals. Always check how your language handles division!

Division by Zero - The Universal Error

All programming languages agree on one thing: you can't divide by zero:

python
```
result = 10 / 0     # Error! Division by zero
```

Always check for zero before dividing:

python
```
divisor = int(input("Enter a number: "))
if divisor != 0:
    result = 100 / divisor
    print(f"Result: {result}")
else:
    print("Cannot divide by zero!")
```

The Modulo Operator (%) - The Remainder Finder

The modulo operator % is probably the most confusing arithmetic operator for beginners, but it's incredibly useful once you understand it.

What Modulo Does

Modulo gives you the remainder after division:

python
```
10 % 3     # 1 (10 ÷ 3 = 3 remainder 1)
15 % 4     # 3 (15 ÷ 4 = 3 remainder 3)
20 % 5     # 0 (20 ÷ 5 = 4 remainder 0)
7 % 2      # 1 (7 ÷ 2 = 3 remainder 1)
8 % 2      # 0 (8 ÷ 2 = 4 remainder 0)
```

Think of it as asking: "If I divide these numbers, what's left over?"

Why Modulo Is Incredibly Useful

Checking if a number is even or odd:

python
```python
number = 17
if number % 2 == 0:
    print("Even number")
else:
    print("Odd number")  # This runs (17 % 2 = 1)
```

Creating cycles and patterns:

python
```python
# Print numbers 0-4 repeatedly
for i in range(20):
    cycle_position = i % 5
    print(cycle_position)  # 0, 1, 2, 3, 4, 0, 1, 2, 3, 4...
```

Time calculations:

python
```python
total_minutes = 90
hours = total_minutes // 60    # 1 (we'll cover // next)
minutes = total_minutes % 60   # 30
print(f"{hours} hours and {minutes} minutes")  # "1 hours and 30 minutes"
```

Checking divisibility:

python
```python
year = 2024
if year % 4 == 0:
    print("Leap year candidate")  # Years divisible by 4
```

Advanced Arithmetic Operators

Now let's look at some operators that might be new to you but are incredibly handy.

Floor Division (//) - Division Without Decimals

Floor division gives you just the whole number part of division, discarding any remainder:

129

python

```python
result = 10 // 3     # 3 (not 3.33...)
result = 15 // 4     # 3 (not 3.75)
result = 20 // 6     # 3 (not 3.33...)
result = 21 // 7     # 3 (exact division still works)
```

Think of // as "division, but round down to the nearest whole number."

Practical use cases:

python

```python
# How many full dozens in 50 items?
items = 50
dozens = items // 12  # 4 dozens

# How many full weeks in 100 days?
days = 100
weeks = days // 7     # 14 weeks

# Converting seconds to minutes (whole minutes only)
seconds = 150
minutes = seconds // 60  # 2 minutes
```

Exponentiation (**) - Raising to a Power

The double asterisk ** raises a number to a power:

python

```python
result = 2 ** 3      # 8 (2 to the power of 3)
result = 5 ** 2      # 25 (5 squared)
result = 10 ** 4     # 10000 (10 to the power of 4)
result = 4 ** 0.5    # 2.0 (square root of 4)
```

Practical examples:

python

```python
# Area of a square
side = 5
area = side ** 2     # 25
```

```python
# Compound interest calculation
principal = 1000
rate = 1.05        # 5% interest
years = 3
amount = principal * (rate ** years)  # 1157.625

# Powers of 2 (common in computing)
memory_sizes = [2 ** i for i in range(10)]  # [1, 2, 4, 8, 16, 32, 64, 128, 256, 512]
```

Order of Operations (PEMDAS/BODMAS)

Programming languages follow the same order of operations as math:

1. **Parentheses** (or brackets)
2. **Exponents** (`**`)
3. **Multiplication and Division** (`*`, `/`, `//`, `%`) - left to right
4. **Addition and Subtraction** (`+`, `-`) - left to right

Examples of Order Matters

python
```python
result = 2 + 3 * 4       # 14 (not 20) - multiply first
result = (2 + 3) * 4     # 20 - parentheses change the order
result = 2 ** 3 * 4      # 32 (not 4096) - exponent first
result = 2 * 3 + 4 * 5   # 26 (not 20) - left to right but same precedence
```

When in Doubt, Use Parentheses

python
```python
# Unclear intention
total = base + tax_rate * base

# Clear intention - tax calculated first
total = base + (tax_rate * base)

# Clear intention - everything added first
```

```
total = (base + tax_rate) * base
```

Increment and Decrement Shortcuts

Many languages offer shortcuts for adding or subtracting 1 from a variable.

The ++ and -- Operators (JavaScript, Java, C++)

javascript

```javascript
// JavaScript examples
let count = 5;

count++;      // count becomes 6 (same as count = count + 1)
count--;      // count becomes 5 (same as count = count - 1)

// These can be used in expressions
let a = 10;
let b = a++;   // b gets 10, then a becomes 11
let c = ++a;   // a becomes 12, then c gets 12
```

Python's Approach (No ++ or --)

Python doesn't have ++ and --, but it has compound assignment operators:

python

```python
count = 5
count += 1     # count becomes 6 (same as count = count + 1)
count -= 1     # count becomes 5 (same as count = count - 1)
```

Compound Assignment Operators

These operators combine arithmetic with assignment to make code shorter:

python

```python
# Basic assignment
score = 100

# Instead of: score = score + 10
score += 10    # score becomes 110
```

```
# Instead of score = score - 5
score -= 5     # score becomes 105

# Instead of score = score * 2
score *= 2     # score becomes 210

# Instead of score = score / 2
score /= 2     # score becomes 105.0

# Works with strings too!
message = "Hello"
message += " World"  # message becomes "Hello World"
```

All the Compound Operators

python
```
x = 10

x += 5    # x = x + 5 --> x becomes 15
x -= 3    # x = x - 3 --> x becomes 12
x *= 2    # x = x * 2 --> x becomes 24
x /= 4    # x = x / 4 --> x becomes 6.0
x //= 2   # x = x // 2 --> x becomes 3.0
x %= 2    # x = x % 2 --> x becomes 1.0
x **= 3   # x = x ** 3 --> x becomes 1.0
```

Common Arithmetic Mistakes

Mistake 1: Integer Division Surprise

python
```
# In some languages or contexts
average = (score1 + score2 + score3) / 3

# If all scores are integers, you might get integer division
# Solution: make sure at least one value is a decimal
average = (score1 + score2 + score3) / 3.0
```

133

Mistake 2: Modulo with Negative Numbers

python

```
# Modulo with negative numbers can be surprising
print(-7 % 3)   # Might be 2 in Python, -1 in other languages
print(7 % -3)   # Results vary by language


# When in doubt, test with your specific language
```

Mistake 3: Floating Point Precision

python

```
# Floating point math isn't always exact
result = 0.1 + 0.2
print(result)   # Might print 0.30000000000000004


# For exact decimal math, use decimal module in Python
from decimal import Decimal
result = Decimal('0.1') + Decimal('0.2')   # Exactly 0.3
```

Mistake 4: Forgetting Order of Operations

python

```
# Intended: calculate 20% tip on $50 bill
tip = 50 * 0.20 + 50    # Wrong! This is (50 * 0.20) + 50 = 60


# Correct: calculate tip, then add to bill
tip = (50 * 0.20) + 50  # Still wrong - this is tip + bill
total = 50 + (50 * 0.20) # Correct - bill + tip = 60


# Or even clearer
bill = 50
tip_amount = bill * 0.20
total = bill + tip_amount
```

Practical Exercises

Exercise 1: Simple Calculator

Create a calculator that performs basic operations:

python

```python
# Get numbers from user
num1 = float(input("Enter first number: "))
num2 = float(input("Enter second number: "))
operation = input("Enter operation (+, -, *, /): ")

# Your code here - perform the calculation
if operation == "+":
    result = num1 + num2
elif operation == "-":
    result = num1 - num2
elif operation == "*":
    result = num1 * num2
elif operation == "/":
    if num2 != 0:
        result = num1 / num2
    else:
        result = "Cannot divide by zero!"
else:
    result = "Invalid operation!"

print(f"Result: {result}")
```

Exercise 2: Time Converter

Convert seconds into hours, minutes, and seconds:

python

```python
total_seconds = 3661  # Example: 1 hour, 1 minute, 1 second

# Your code here - break down into hours, minutes, seconds
hours = total_seconds // 3600
remaining_seconds = total_seconds % 3600
minutes = remaining_seconds // 60
seconds = remaining_seconds % 60
```

```python
print(f"{total_seconds} seconds = {hours}h {minutes}m {seconds}s")
# Should print: 3661 seconds = 1h 1m 1s
```

Exercise 3: Even/Odd Checker

Check if numbers are even or odd:

python
```python
numbers = [12, 17, 20, 33, 44, 51]

# Your code here - check each number
for number in numbers:
    if number % 2 == 0:
        print(f"{number} is even")
    else:
        print(f"{number} is odd")
```

Exercise 4: Compound Interest Calculator

Calculate compound interest:

python
```python
principal = 1000    # Starting amount
rate = 0.05         # 5% annual interest
years = 10          # 10 years

# Your code here - calculate final amount
# Formula: A = P(1 + r)^t
final_amount = principal * (1 + rate) ** years
interest_earned = final_amount - principal

print(f"Initial: ${principal}")
print(f"After {years} years: ${final_amount:.2f}")
print(f"Interest earned: ${interest_earned:.2f}")
```

Arithmetic in Different Languages

While the concepts are universal, syntax can vary:

Python:

```python
python
result = 10 / 3    # 3.33... (always decimal)
result = 10 // 3   # 3 (floor division)
result = 2 ** 8    # 256 (exponentiation)
```

JavaScript:

```javascript
javascript
let result = 10 / 3;            // 3.33... (always decimal)
let result = Math.floor(10 / 3); // 3 (manual floor)
let result = 2 ** 8;           // 256 (exponentiation)
let result = Math.pow(2, 8);   // 256 (alternative)
```

Java:

```java
java
int result = 10 / 3;          // 3 (integer division)
double result = 10.0 / 3;     // 3.33... (decimal division)
double result = Math.pow(2, 8); // 256.0 (exponentiation)
```

Key Takeaways

Before we move to assignment operators, here are the essential points about arithmetic operators:

1. **Basic operators** (+, -, *, /) work mostly like regular math, but watch for type differences
2. **Division behavior varies** by language - some give integers, others give decimals
3. **Modulo (%) gives remainders** - super useful for cycles, even/odd checks, and time calculations
4. **Floor division (//)** gives whole numbers only - great for "how many complete groups" questions
5. **Exponentiation (**)** raises numbers to powers - handy for compound calculations
6. **Order of operations matters** - use parentheses when in doubt
7. **Compound operators** (+=, -=, etc.) are shortcuts for modifying variables
8. **Always check for division by zero** before dividing

Understanding arithmetic operators gives your programs the ability to calculate, measure, and quantify anything. From simple shopping cart totals to complex scientific calculations, these operators are the foundation of computational thinking.

In the next chapter, we'll explore assignment operators in more detail, including the powerful compound assignment operators that can make your code cleaner and more efficient.

Chapter 7 — Assignment & Compound Operators

Chapter 7 — Assignment & Compound Operators Basic Assignment = The assignment operator = stores values in variables, forming the backbone of data management in code. [Note: Add a visual: e.g., a diagram showing x = 10 with an arrow to a variable box.] [Rest of chapter content: Replace with your original text for "Shortcuts: +=, -=, *=, /=" section.]

Basic Assignment (=) - The Storage Manager

The single equals sign is the most fundamental operator in programming. It doesn't compare values—it stores them.

How Assignment Really Works

When you write `name = "Alice"`, you're not saying "name equals Alice." You're saying "put the value Alice into the container called name."

python
```
# Creating and storing values
age = 25
name = "Alice"
is_student = True
height = 5.8


# The right side is calculated first, then stored
score = 85 + 15      # Calculate 100, then store it in score
full_name = "Alice" + " " + "Johnson"  # Calculate "Alice Johnson", then store it
```

Assignment Is Right-to-Left

This is crucial to understand: assignment always works from right to left. The computer:

1. Calculates everything on the right side of the =
2. Takes that final result

138

3. Stores it in the variable on the left side

python

```
# Step by step breakdown
x = 5 * 3 + 2
# Step 1: Calculate 5 * 3 = 15
# Step 2: Calculate 15 + 2 = 17
# Step 3: Store 17 in variable x


# The variable name doesn't affect the calculation
result = 10 / 2      # Calculates to 5.0, stores in result
answer = 10 / 2      # Same calculation, same result, different variable
```

Reassignment - Variables Can Change

Unlike math, where x = 5 means x is always 5, programming variables can be reassigned:

python

```
count = 0          # count starts at 0
count = count + 1     # count becomes 1
count = count + 1     # count becomes 2
count = 100        # count is now 100 (previous value is gone)
```

This is perfectly normal and incredibly useful:

python

```
# Running total
total = 0
total = total + 50     # Add $50
total = total + 25     # Add $25 (total is now $75)
total = total - 10     # Subtract $10 (total is now $65)


# User input processing
user_input = input("Enter your name: ")
user_input = user_input.strip()     # Remove extra spaces
user_input = user_input.capitalize() # Capitalize first letter
print(f"Hello, {user_input}!")
```

The Assignment vs. Equality Confusion

This is probably the #1 source of confusion for new programmers:

python

```python
# ASSIGNMENT (=) - stores a value
age = 18        # Put 18 into the variable age

# COMPARISON (==) - checks if values are equal
if age == 18:        # Check if age contains 18
    print("You're 18!")
```

The Dangerous Mix-Up

python

```python
# Wrong - this assigns 18 to age (and might always be true)
if age = 18:
    print("This might not work as expected")

# Right - this compares age to 18
if age == 18:
    print("You're exactly 18 years old")
```

In Python, using = in an if-statement will cause an error (thankfully). But in some languages like C or JavaScript, it's valid code that does something completely different from what you intended.

Compound Assignment Operators - The Shortcuts

Compound assignment operators combine arithmetic with assignment. They're shortcuts that make common patterns much cleaner.

The += Operator - Add and Assign

Instead of writing x = x + 5, you can write x += 5:

python

```python
# The long way
score = 100
score = score + 10    # score becomes 110

# The shortcut way
```

140

```python
score = 100
score += 10        # score becomes 110 (same result)
```

Real-world examples:

python
```python
# Running totals
total_price = 0
total_price += 19.99  # Add item 1
total_price += 5.50   # Add item 2
total_price += 12.25  # Add item 3
print(f"Total: ${total_price}")  # Total: $37.74


# Accumulating points
player_score = 0
player_score += 100   # Beat level 1
player_score += 250   # Beat level 2
player_score += 500   # Beat boss
print(f"Final score: {player_score}")  # Final score: 850
# Building strings
message = "Hello"
message += ", "        # "Hello, "
message += "World"     # "Hello, World"
message += "!"         # "Hello, World!"
```

The -= Operator - Subtract and Assign

python
```python
# Reducing values
health = 100
health -= 25           # Take 25 damage (health becomes 75)
health -= 10           # Take 10 more damage (health becomes 65)


# Countdown
countdown = 10
countdown -= 1         # 9
countdown -= 1         # 8
countdown -= 1         # 7
```

141

The *= Operator - Multiply and Assign

python
```
# Doubling values
money = 100
money *= 2        # money becomes 200

# Applying discounts (multiply by percentage)
price = 50.00
price *= 0.9       # Apply 10% discount (price becomes 45.00)

# Compound interest
balance = 1000
balance *= 1.05     # 5% interest (balance becomes 1050.00)

# String repetition
decoration = "="
decoration *= 20    # decoration becomes "===================="
```

The /= Operator - Divide and Assign

python
```
# Splitting evenly
pizza_slices = 8
pizza_slices /= 2    # pizza_slices becomes 4.0

# Converting units
meters = 1000
meters /= 100       # Convert to centimeters (meters becomes 10.0)

# Calculating averages (step by step)
total_scores = 285
total_scores /= 3   # total_scores becomes 95.0 (average)
```

The //= Operator - Floor Divide and Assign

python
```
# Integer division
```

```
items = 17
items //= 5        # items becomes 3 (17 // 5 = 3, remainder ignored)


# Time conversion
total_minutes = 150
total_minutes //= 60   # total_minutes becomes 2 (hours)
```

The %= Operator - Modulo and Assign

python
```
# Wrapping numbers in ranges
position = 12
position %= 10       # position becomes 2 (12 % 10 = 2)


# Creating cycles
day_of_week = 15
day_of_week %= 7     # day_of_week becomes 1 (Monday, if Sunday is 0)


# Keeping numbers within bounds
level = 25
level %= 5           # level becomes 0 (25 % 5 = 0)
```

The **= Operator - Exponent and Assign

python
```
# Squaring numbers
side_length = 5
side_length **= 2    # side_length becomes 25 (5 squared)


# Compound growth
population = 1000
population **= 1.1    # population becomes 1000 ^ 1.1 = 1096
```

Multiple Assignment - Assigning Several Variables at Once

Some languages let you assign multiple variables in one line:

Tuple Assignment (Python)

python

```python
# Assign multiple variables at once
name, age, city = "Alice", 25, "New York"
# Same as:
# name = "Alice"
# age = 25
# city = "New York"

# Swapping variables (the Python way)
a = 5
b = 10
a, b = b, a        # Now a is 10 and b is 5!

# Multiple function returns
def get_name_age():
    return "Bob", 30


person_name, person_age = get_name_age()
```

Chained Assignment

python

```python
# Assign the same value to multiple variables
x = y = z = 0        # All three variables become 0
# Same as:
# x = 0
# y = 0
# z = 0

# Useful for initialization
player1_score = player2_score = player3_score = 0
red = green = blue = 255  # White color in RGB
```

Be careful with mutable objects:

python

```python
# This creates separate lists (good)
list1 = []
```

```python
list2 = []
list3 = []

# This creates one list shared by all variables (usually not what you want)
list1 = list2 = list3 = []
list1.append("item")
print(list2)  # Also contains "item"?
```

Augmented Assignment with Different Data Types

Compound assignment operators work with more than just numbers:

String Concatenation

python
```python
message = "Hello"
message += " there"    # "Hello there"
message += "!"         # "Hello there!"

# Building file paths
path = "/home/user"
path += "/documents"   # "/home/user/documents"
path += "/file.txt"    # "/home/user/documents/file.txt"

# Creating formatted output
output = "Name: "
output += user_name
output += ", Age: "
output += str(user_age)
```

List Operations

python
```python
# Adding items to lists
shopping_list = ["milk", "bread"]
shopping_list += ["eggs"]              # ["milk", "bread", "eggs"]
shopping_list += ["cheese", "butter"]  # ["milk", "bread", "eggs", "cheese", "butter"]

# Multiplying lists
```

145

```python
numbers = [1, 2, 3]
numbers *= 2        # [1, 2, 3, 1, 2, 3]
```

Common Assignment Mistakes

Mistake 1: Confusing = and ==

python
```python
# Wrong - assigns value instead of comparing
user_age = 18
if user_age = 21:     # This is assignment, not comparison!
    print("Can drink")

# Right - compares values
if user_age == 21:
    print("Can drink")
```

Mistake 2: Using Compound Operators on Uninitialized Variables

python
```python
# Wrong - what's the starting value of total?
total += 50        # Error! total doesn't exist yet

# Right - initialize first
total = 0
total += 50        # Now total is 50
```

Mistake 3: Forgetting That Assignment Returns a Value

python
```python
# In some languages, this is valid but confusing
a = b = c = 5        # All get 5
a = (b = c + 1)      # Confusing! Avoid this style

# Clearer approach
c = 5
```

146

```
b = c + 1
a = b
```

Mistake 4: Modifying the Wrong Variable

python
```
# Intended: increase the score
player_score = 100
total_score = 500


# Oops! Modified the wrong variable
total_score += 50      # total_score becomes 550
print(player_score)    # Still 100 - we didn't change it!


# Correct
player_score += 50     # player_score becomes 150
```

When to Use Each Assignment Operator

Use Basic Assignment (=) When:

- Setting initial values
- Storing calculation results
- Assigning completely new values
- Working with complex expressions

python
```
# Initial setup
name = "Alice"
starting_balance = 1000.00


# Calculation results
area = length * width
final_grade = (exam1 + exam2 + project) / 3


# Complex expressions
```

```
formatted_name = user_input.strip().title().replace(" ", "_")
```

Use Compound Assignment (+=, -=, etc.) When:

- Modifying existing values
- Building running totals
- Incrementing counters
- Accumulating results

python
```
# Modifying existing values
score += bonus_points
temperature -= 10


# Running totals
cart_total += item_price
error_count += 1


# Accumulating results
log_message += f"Error at line {line_number}\n"
```

Practical Exercises

Exercise 1: Shopping Cart Calculator

Build a simple shopping cart that uses various assignment operators:

python
```
# Start with empty cart
cart_total = 0.00
item_count = 0


# Add items using compound assignment
print("Adding items to cart...")
cart_total += 19.99    # Add shirt
item_count += 1


cart_total += 45.50    # Add shoes
item_count += 1
```

```python
cart_total += 12.75      # Add socks
item_count += 1


# Apply discount (10% off)
print(f"Subtotal: ${cart_total}")
cart_total *= 0.9        # 10% discount
print(f"After discount: ${cart_total:.2f}")


# Add tax (8.5%)
cart_total *= 1.085      # Add tax
print(f"Final total: ${cart_total:.2f}")
print(f"Total items: {item_count}")
```

Exercise 2: Game Score Tracker

Track a player's score through multiple levels:

python
```python
# Initialize player stats
player_score = 0
lives = 3
level = 1


# Level 1
print(f"Level {level}")
player_score += 100      # Base points
player_score += 50       # Bonus for speed
print(f"Score: {player_score}")


# Level 2 - player takes damage
level += 1
print(f"Level {level}")
player_score += 200      # Base points
lives -= 1               # Hit by enemy
player_score -= 25       # Penalty for damage
print(f"Score: {player_score}, Lives: {lives}")


# Level 3 - player reaches bonus score
```

149

```python
level += 1
print(f"Level {level}")
player_score += 300    # Base points
player_score *= 2      # Power-up double
print(f"Final Score: {player_score}")
```

Exercise 3: Text Message Builder

Build a formatted message using string assignment:

python
```python
# Start with empty message
message = ""
sender = "Alice"
recipient = "Bob"
timestamp = "2:30 PM"

# Build the message
message += "From: "
message += sender
message += "\nTo: "
message += recipient
message += "\nTime: "
message += timestamp
message += "\n" + "="*20 + "\n"
message += "Hey Bob, want to grab lunch today?\n"
message += "Let me know if you're free!\n"
message += "="*20

print(message)
```

Assignment in Different Languages

The concepts are universal, but syntax varies:

Python:

python
```python
x = 10
```

```
x += 5        # x becomes 15
name = "Alice"
name += " Smith"  # "Alice Smith"
```

JavaScript:

```javascript
javascript
let x = 10;
x += 5;        // x becomes 15
let name = "Alice";
name += " Smith"; // "Alice Smith"
```

Java:

```java
java
int x = 10;
x += 5;        // x becomes 15
String name = "Alice";
name += " Smith"; // "Alice Smith"
```

Key Takeaways

Before we move to logical operators, here are the essential points about assignment operators:

1. **Single equals (=) assigns values** - it doesn't compare them
2. **Assignment works right-to-left** - calculate the right side first, then store
3. **Variables can be reassigned** - the old value is replaced with the new one
4. **Compound operators** (+=, -=, *=, etc.) are shortcuts for common patterns
5. **Always initialize variables** before using compound assignment
6. **Multiple assignment** can set several variables at once
7. **Be careful with shared objects** in chained assignment
8. **Use basic assignment for new values**, compound assignment for modifications

Understanding assignment operators is crucial because they're how your program remembers things. Every time you store user input, save a calculation result, or update a running total, you're using assignment operators.

In the next chapter, we'll explore logical operators—the symbols that help your programs make complex decisions by combining multiple conditions with "and," "or," and "not."

Chapter 8: Arrow Operators and Modern Symbols

The Evolution of Programming Symbols

Programming languages are living things that evolve over time. As programmers discovered better ways to express complex ideas, new symbols emerged. Arrow operators are some of the newest additions to programming languages, and they represent a shift toward more expressive, functional programming styles.

When I first saw symbols like ->, =>, and ::, I was completely baffled. They looked like emoticons or ASCII art, not legitimate programming symbols. But once I learned what they do, I realized they make code much more readable and elegant. These modern symbols let you write in a style that's closer to mathematical notation and natural language.

In this chapter, we'll explore these modern symbols, understand their different meanings across languages, and learn when and how to use them to write cleaner, more expressive code.

Arrow Functions (=>) - The Lambda Makers

The fat arrow => is probably the most recognizable modern programming symbol. It creates anonymous functions (functions without names) in a concise way.

JavaScript Arrow Functions

javascript

```javascript
// Traditional function
function double(x) {
    return x * 2;
}

// Arrow function - much shorter!
const double = x => x * 2;

// Multiple parameters need parentheses
const add = (a, b) => a + b;

// Complex functions use braces
const processUser = user => {
    console.log(`Processing ${user.name}`);
```

```javascript
  return user.name.toUpperCase();
};
```

Why Arrow Functions Matter

Before arrows (verbose and repetitive):

javascript

```javascript
const numbers = [1, 2, 3, 4, 5];

const doubled = numbers.map(function(num) {
  return num * 2;
});

const evens = numbers.filter(function(num) {
  return num % 2 === 0;
});
```

With arrows (clean and concise):

javascript

```javascript
const numbers = [1, 2, 3, 4, 5];

const doubled = numbers.map(num => num * 2);
const evens = numbers.filter(num => num % 2 === 0);
```

Python Lambda Functions

Python uses `lambda` keyword, but the concept is similar:

python

```python
# Traditional function
def double(x):
  return x * 2

# Lambda function (anonymous)
double = lambda x: x * 2

# Used with map, filter, etc.
numbers = [1, 2, 3, 4, 5]
```

```
doubled = list(map(lambda x: x * 2, numbers))
evens = list(filter(lambda x: x % 2 == 0, numbers))
```

C# Lambda Expressions

csharp
```
// C# uses => for lambda expressions
var numbers = new[] { 1, 2, 3, 4, 5 };


var doubled = numbers.Select(x => x * 2);
var evens = numbers.Where(x => x % 2 == 0);


// More complex lambdas
var users = GetUsers();
var adults = users.Where(user => user.Age >= 18)
        .Select(user => new { user.Name, user.Age });
```

Object Access Arrows (->) - The Pointer Dereferencers

The thin arrow -> has different meanings depending on the programming language, but it's most commonly associated with accessing members through pointers or references.

C++ Pointer Member Access

cpp
```
// C++ - accessing object members through pointers
struct Person {
    string name;
    int age;
    void greet() { cout << "Hello, I'm " << name; }
};


Person* personPtr = new Person{"Alice", 30};


// Using -> to access members through pointer
```

```cpp
cout << personPtr->name;   // "Alice"
cout << personPtr->age;    // 30
personPtr->greet();        // "Hello, I'm Alice"

// Equivalent to (*personPtr).name, but much cleaner
cout << (*personPtr).name; // Same as above, but verbose
```

PHP Object Access

php
```php
// PHP uses -> for object property and method access
class Person {
    public $name;
    public $age;

    public function greet() {
        return "Hello, I'm " . $this->name;
    }
}

$person = new Person();
$person->name = "Alice";
$person->age = 30;
echo $person->greet(); // "Hello, I'm Alice"
```

Rust Pattern Matching

rust
```rust
// Rust uses -> to match expressions and function signatures
fn process_result(result: Result<i32, String>) -> String {
    match result {
        Ok(value) => format!("Success: {}", value),
        Err(error) => format!("Error: {}", error),
    }
}

// Function return type annotation
fn calculate_tax(price: f64) -> f64 {
```

```
    price * 0.08
}
```

Scope Resolution (::) - The Namespace Navigators

The double colon : : helps you access things that are defined in specific namespaces, modules, or classes.

C++ Namespace Access

cpp
```cpp
// C++ namespaces and scope resolution
namespace Math {
    const double PI = 3.14159;
    double square(double x) { return x * x; }
}

namespace Graphics {
    const double PI = 3.14;  // Different PI value
    void drawCircle(double radius) { /* ... */ }
}

// Using :: to specify which PI you want
double area = Math::PI * radius * radius;
Graphics::drawCircle(5.0);

// Global scope resolution
int globalVar = 100;
void someFunction() {
    int globalVar = 50;      // Local variable shadows global
    cout << ::globalVar;     // Access global version (100)
    cout << globalVar;       // Access local version (50)
}
```

Rust Module Access

rust
```rust
// Rust modules and paths
mod math {
```

```rust
    pub const PI: f64 = 3.14159;
    pub fn square(x: f64) -> f64 { x * x }
}

mod graphics {
    pub const PI: f64 = 3.14;
    pub fn draw_circle(radius: f64) { /* ... */ }
}

// Using :: to access module items
let area = math::PI * radius * radius;
graphics::draw_circle(5.0);

// Standard library access
let numbers = vec![1, 2, 3];
let max_value = std::cmp::max(10, 20);
```

Python Class Methods and Static Access

python

```python
# Python uses : in some contexts (like type hints)
from typing import List, Dict

# Class methods and static methods
class MathUtils:
    PI = 3.14159

    @staticmethod
    def square(x):
        return x * x

    @classmethod
    def circle_area(cls, radius):
        return cls.PI * radius * radius
```

```
# Accessing static members
area = MathUtils.circle_area(5)
squared = MathUtils.square(4)
```

Pipeline Operators (|>) - The Data Flow Directors

The pipeline operator is newer and not available in all languages, but it's becoming popular for making data transformations more readable.

F# Pipelines

fsharp
```
// F# pipeline operator - data flows left to right
let numbers = [1; 2; 3; 4; 5]

let result = numbers
        |> List.filter (fun x -> x % 2 = 0)
        |> List.map (fun x -> x * x)
        |> List.sum

// Same as: List.sum(List.map(fun x -> x * x, List.filter(fun x -> x % 2 = 0, numbers)))
// Much more readable with pipelines!
```

Elixir Pipelines

elixir
```
# Elixir - pipeline operator is central to the language
"hello world"
|> String.split()
|> Enum.map(&String.capitalize/1)
|> Enum.join(" ")
# Result: "Hello World"

# Without pipelines (nested and hard to read):
# Enum.join(Enum.map(String.split("hello world"), &String.capitalize/1), " ")
```

Proposed JavaScript Pipeline

javascript

```javascript
const result = value
  |> doSomething
  |> doSomethingElse
  |> finalize;
```

```javascript
const result = finalize(doSomethingElse(doSomething(value)));
```

Nullish Coalescing (??) - The Default Providers

The nullish coalescing operator provides default values when dealing with null or undefined values.

JavaScript Nullish Coalescing

javascript

```javascript
// ?? provides defaults for null or undefined (but not other falsy values)
let username = user.name ?? "Anonymous";
let port = config.port ?? 3000;
let theme = settings.theme ?? "light";

// Different from || operator
let count1 = 0;
let result1 = count1 || 10;    // 10 (it is false)
let result2 = count1 ?? 10;    // 0 (it is not null/undefined)

// Chaining for nested properties
let city = user?.profile?.address?.city ?? "Unknown";
```

C# Null Coalescing

csharp

```csharp
// C# has had ?? for much longer
string username = user.Name ?? "Anonymous";
int port = config.Port ?? 3000;

// Null coalescing assignment (??=)
username ??= "DefaultUser"; // Only assign if username is null
```

Optional Chaining (?.) - The Safe Accessors

Optional chaining lets you safely access nested properties without worrying about null/undefined errors.

JavaScript Optional Chaining

javascript

```javascript
// Without optional chaining (verbose and error-prone)
let city;
if (user && user.profile && user.profile.address) {
  city = user.profile.address.city;
} else {
  city = undefined;
}


// With optional chaining (clean and safe)
let city = user?.profile?.address?.city;


// Works with methods too
let result = api?.getData?.();


// Array access
let firstItem = items?.[0];
let dynamicProperty = user?.[propertyName];
```

Swift Optional Chaining

swift

```swift
// Swift has similar optional chaining
class Person {
  var address: Address?
}


class Address {
  var city: String?
}
```

160

```javascript
let person = Person()
let city = person.address?.city  // Returns nil if address is nil
```

Spread and Rest Operators (...) - The Collectors and Distributors

The three dots . . . have multiple uses in modern programming languages.

JavaScript Spread Operator

javascript
```javascript
// Spread arrays
let numbers1 = [1, 2, 3];
let numbers2 = [4, 5, 6];
let combined = [...numbers1, ...numbers2];  // [1, 2, 3, 4, 5, 6]

// Spread objects
let person = { name: "Alice", age: 30 };
let employee = { ...person, job: "Developer" };
// { name: "Alice", age: 30, job: "Developer" }

// Function calls
function sum(a, b, c) {
    return a + b + c;
}
let nums = [1, 2, 3];
let total = sum(...nums);  // Same as sum(1, 2, 3)
```

JavaScript Rest Parameters

javascript
```javascript
// Rest parameters - collect remaining arguments
function sum(first, ...rest) {
    console.log("First:", first);  // 1
    console.log("Rest:", rest);    // [2, 3, 4, 5]
    return rest.reduce((acc, num) => acc + num, first);
}

sum(1, 2, 3, 4, 5);
```

```javascript
// Destructuring with rest
let [first, second, ...remaining] = [1, 2, 3, 4, 5];
console.log(first);      // 1
console.log(second);     // 2
console.log(remaining);  // [3, 4, 5]
```

Python Unpacking

python
```python
# Python uses * for similar functionality
numbers1 = [1, 2, 3]
numbers2 = [4, 5, 6]
combined = [*numbers1, *numbers2]  # [1, 2, 3, 4, 5, 6]

# Function arguments
def greet(name, *args, **kwargs):
    print(f"Hello {name}")
    print(f"Additional args: {args}")
    print(f"Keyword args: {kwargs}")

greet("Alice", "extra", "info", age=30, city="NYC")
```

Template Literals and String Interpolation

Modern languages provide better ways to build strings with embedded values.

JavaScript Template Literals

javascript
```javascript
// Template literals with backticks
let name = "Alice";
let age = 30;
let message = `Hello, my name is ${name} and I am ${age} years old.`;

// Multi-line strings
let html = `
<div class="user">
    <h2>${name}</h2>
    <p>Age: ${age}</p>
```

162

```
</div>
`;
```

```
// Expression evaluation
let price = 19.99;
let tax = 0.08;
let receipt = `Total: $${(price * (1 + tax)).toFixed(2)}`;
```

Python f-strings

python
```
# Python f-strings (Python 3.6+)
name = "Alice"
age = 30
message = f"Hello, my name is {name} and I am {age} years old."

# Expression evaluation
price = 19.99
tax = 0.08
receipt = f"Total: ${price * (1 + tax):.2f}"

# Multi-line f-strings
html = f"""
<div class="user">
    <h2>{name}</h2>
    <p>Age: {age}</p>
</div>
"""
```

Practical Exercises

Exercise 1: Arrow Function Conversion

javascript
```
// Convert these traditional functions to arrow functions
function square(x) {
    return x * x;
}
```

```javascript
function isEven(num) {
  return num % 2 === 0;
}


function greetUser(name) {
  console.log(`Hello, ${name}!`);
  return `Greeting sent to ${name}`;
}


// Solutions:
const square = x => x * x;
const isEven = num => num % 2 === 0;
const greetUser = name => {
  console.log(`Hello, ${name}!`);
  return `Greeting sent to ${name}`;
};


// Test them
console.log(square(5));      // 25
console.log(isEven(4));      // true
console.log(greetUser("Alice")); // "Greeting sent to Alice"
```

Exercise 2: Safe Property Access

javascript
```javascript
// Practice with optional chaining and nullish coalescing
let users = [
  {
    name: "Alice",
    profile: {
      address: {
        city: "New York",
        zip: "10001"
      }
    }
  },
```

```javascript
  {
    name: "Bob",
    profile: {
      // No address
    }
  },
  {
    name: "Charlie"
    // No profile
  }
];

// Your task: safely extract city information for each user
// Use optional chaining (?) and nullish coalescing (??)

users.forEach(user => {
  let city = user?.profile?.address?.city ?? "City Unknown";
  let zip = user?.profile?.address?.zip ?? "ZIP Unknown";

  console.log(`${user.name}: ${city}, ${zip}`);
});

// Expected output:
// Alice: New York, 10001
// Bob: City Unknown, ZIP Unknown
// Charlie: City Unknown, ZIP Unknown
```

Exercise 3: Data Processing Pipeline

javascript

```javascript
// Create a data processing pipeline using modern operators
let salesData = [
  { product: "Laptop", price: 999, quantity: 2, category: "Electronics" },
  { product: "Mouse", price: 25, quantity: 5, category: "Electronics" },
  { product: "Book", price: 15, quantity: 3, category: "Books" },
  { product: "Phone", price: 699, quantity: 1, category: "Electronics" },
  { product: "Notebook", price: 5, quantity: 10, category: "Books" }
```

```javascript
];

// Your task: Use arrow functions and array methods to:
// 1. Calculate total value for each item (price * quantity)
// 2. Filter for items worth more than $50
// 3. Group by category
// 4. Calculate category totals

// Solution:
const processedData = salesData
  .map(item => ({
    ...item,
    totalValue: item.price * item.quantity
  }))
  .filter(item => item.totalValue > 50)
  .reduce((acc, item) => {
    const category = item.category;
    if (!acc[category]) {
      acc[category] = { items: [], total: 0 };
    }
    acc[category].items.push(item);
    acc[category].total += item.totalValue;
    return acc;
  }, {});

console.log(processedData);
```

Modern Symbols in Different Languages

JavaScript:

```javascript
javascript
const double = x => x * 2;          // Arrow function
const city = user?.address?.city;   // Optional chaining
const name = user.name ?? "Unknown"; // Nullish coalescing
const combined = [...arr1, ...arr2]; // Spread operator
```

C++:

166

```cpp
cpp
person->getName();          // Pointer member access
Math::PI;                   // Namespace access
auto lambda = [](int x) { return x * 2; }; // Lambda function
```

Python:

```python
python
result = lambda x: x * 2      # Lambda function
message = f"Hello {name}!"     # f-string interpolation
combined = [*list1, *list2]    # Unpacking
```

Rust:

```rust
rust
let result = numbers.iter()      // Method chaining
    .filter(|&x| x > 0)          // Closure
    .map(|&x| x * 2)             // Another closure
    .collect::<Vec<_>>();        // Turbofish operator ::<>
```

Key Takeaways

Modern programming symbols make code more expressive and concise:

1. **Arrow functions (=>)** create anonymous functions with cleaner syntax
2. **Object access arrows (->)** dereference pointers and access object members
3. **Scope resolution (::)** navigates namespaces and modules
4. **Pipeline operators (|>)** make data transformations more readable
5. **Nullish coalescing (??)** provides safe default values
6. **Optional chaining (?.)** prevents null/undefined errors
7. **Spread/Rest (...)** collect and distribute array/object elements
8. **Template literals** enable string interpolation with embedded expressions

These modern symbols represent the evolution of programming languages toward more functional, expressive styles. While they might seem intimidating at first, they actually make code more readable and less prone to errors once you understand them.

Understanding these symbols prepares you for modern codebases and helps you write cleaner, more maintainable code. They're not just syntactic sugar—they represent fundamental shifts in how we think about programming problems.

Chapter 9: Language-Specific Symbols and Operators

The Unique Flavors of Programming Languages

While most programming symbols are universal, each language has its own special operators that make it unique. These language-specific symbols often represent the philosophy and design goals of the language—Python's emphasis on readability, JavaScript's flexibility with types, Java's strict object-oriented approach, and C++'s low-level control.

When I started learning multiple programming languages, these unique symbols were initially frustrating. I'd see `instanceof` in Java and wonder why Python didn't have it, or encounter Python's `is` operator and feel lost when switching to JavaScript. But I eventually realized these differences aren't bugs—they're features that make each language suited for different types of problems.

In this chapter, we'll explore the distinctive symbols and operators that give each major programming language its personality. Understanding these differences will make you a more versatile programmer and help you choose the right language for each project.

Python's Distinctive Operators

Python prides itself on readability and has several operators that use English words instead of cryptic symbols.

The `in` Operator - Membership Testing

Python's `in` operator checks if something is contained within something else:

python

```python
# Lists and strings
fruits = ["apple", "banana", "orange"]
if "apple" in fruits:
    print("We have apples!")  # This runs

# String contains
text = "Hello, World!"
if "World" in text:
    print("Found 'World' in the text")  # This runs

# Dictionary keys
person = {"name": "Alice", "age": 30}
if "name" in person:
    print("Person has a name")  # This runs
```

168

```python
# Range membership
if 5 in range(1, 10):
    print("5 is between 1 and 10")  # This runs
```

Practical applications:

python

```python
# Input validation
valid_colors = ["red", "green", "blue", "yellow"]
user_color = input("Choose a color: ")
if user_color.lower() in valid_colors:
    print(f"Great choice: {user_color}")
else:
    print("Please choose red, green, blue, or yellow")


# File extension checking
filename = "document.pdf"
if filename.endswith(".pdf") or ".pdf" in filename:
    print("This is a PDF file")


# Substring search
email = "user@example.com"
if "@" in email and "." in email:
    print("Valid email format")
```

The `is` Operator - Identity Testing

While == checks if values are equal, `is` checks if they're the exact same object in memory:

python

```python
# Comparing values vs identity
a = [1, 2, 3]
b = [1, 2, 3]
c = a

print(a == b)   # True (same values)
print(a is b)   # False (different objects)
print(a is c)   # True (same object)
```

169

```python
# Common use with None
user_name = None
if user_name is None:
    print("No name provided")  # Preferred over user_name == None

# Boolean identity
flag = True
if flag is True:
    print("Flag is exactly True")

# Be careful with small integers (they're cached!)
x = 5
y = 5
print(x is y)    # True (Python caches small integers)

x = 1000
y = 1000
print(x is y)    # False (larger integers aren't cached)
```

The `not in` and `is not` Combinations

Python lets you combine these operators with `not`:

python
```python
# not in
forbidden_words = ["spam", "scam", "virus"]
message = "This is a legitimate email"
if not any(word in message.lower() for word in forbidden_words):
    print("Message appears clean")

# Cleaner version
if all(word not in message.lower() for word in forbidden_words):
    print("Message appears clean")

# is not
result = get_user_data()
if result is not None:
```

```python
    print(f"Got data: {result}")
else:
    print("No data available")
```

Python's Walrus Operator (:=) - Assignment Expressions

Introduced in Python 3.8, the walrus operator allows assignment within expressions:

python
```python
# Traditional way - variable used once
user_input = input("Enter a number: ")
if len(user_input) > 0:
    number = int(user_input)
    print(f"You entered: {number}")


# With walrus operator - more concise
if (user_input := input("Enter a number: ")) and len(user_input) > 0:
    number = int(user_input)
    print(f"You entered: {number}")


# File processing example
import re
text = "The price is $29.99 and tax is $2.40"
if (match := re.search(r'\$(\d+\.\d+)', text)):
    price = float(match.group(1))
    print(f"Found price: ${price}")


# List comprehension with walrus
numbers = [1, 2, 3, 4, 5, 6, 7, 8, 9, 10]
squared_evens = [square for n in numbers if (square := n * n) % 2 == 0]
print(squared_evens)  # [4, 16, 36, 64, 100]
```

JavaScript's Type-Related Operators

JavaScript's flexible type system requires special operators to work with types safely.

The typeof Operator - Type Checking

javascript

```javascript
// Basic type checking
console.log(typeof 42);          // "number"
console.log(typeof "hello");     // "string"
console.log(typeof true);        // "boolean"
console.log(typeof undefined);   // "undefined"
console.log(typeof null);        // "object" (this is a famous JavaScript quirk!)
console.log(typeof {});          // "object"
console.log(typeof []);          // "object" (arrays are objects in JS)
console.log(typeof function(){}); // "function"

// Practical type checking
function processValue(value) {
  if (typeof value === "string") {
    return value.toUpperCase();
  } else if (typeof value === "number") {
    return value * 2;
  } else if (typeof value === "boolean") {
    return !value;
  } else {
    return "Unknown type";
  }
}

console.log(processValue("hello")); // "HELLO"
console.log(processValue(42));      // 84
console.log(processValue(true));    // false
```

The `instanceof` Operator - Constructor Checking

javascript
```javascript
// Checking object constructors
let date = new Date();
let array = [1, 2, 3];
let regex = /pattern/;

console.log(date instanceof Date);   // true
console.log(array instanceof Array); // true
```

172

```javascript
console.log(regex instanceof RegExp);   // true

// Custom constructors
function Person(name) {
  this.name = name;
}

let alice = new Person("Alice");
console.log(alice instanceof Person);   // true
console.log(alice instanceof Object);   // true (Person inherits from Object)

// Practical use in functions
function formatDate(input) {
  if (input instanceof Date) {
    return input.toLocaleDateString();
  } else if (typeof input === "string") {
    return new Date(input).toLocaleDateString();
  } else {
    throw new Error("Invalid date input");
  }
}
```

The `in` Operator - Property Checking

JavaScript also has an `in` operator, but it works differently than Python's:

javascript

```javascript
// Checking object properties
let person = { name: "Alice", age: 30 };
console.log("name" in person);     // true
console.log("height" in person);   // false

// Inherited properties
console.log("toString" in person); // true (inherited from Object)

// Array indices
let fruits = ["apple", "banana", "orange"];
console.log(0 in fruits);          // true
```

173

```javascript
console.log(3 in fruits);        // false
console.log("length" in fruits);  // true

// Practical property checking
function safeAccess(obj, property) {
  if (property in obj) {
    return obj[property];
  } else {
    return `Property '${property}' not found`;
  }
}
```

Java's Object-Oriented Operators

Java's strict typing system includes operators specifically for object-oriented programming.

The `instanceof` Operator - Type Hierarchy Checking

java

```java
// Basic instanceof usage
String text = "Hello";
Object obj = text;

if (obj instanceof String) {
  System.out.println("It's a string: " + ((String) obj).toUpperCase());
}

// Inheritance hierarchies
class Animal {
  void makeSound() { System.out.println("Some sound"); }
}

class Dog extends Animal {
  void bark() { System.out.println("Woof!"); }
}

class Cat extends Animal {
  void meow() { System.out.println("Meow!"); }
```

174

```java
}

Animal[] animals = {new Dog(), new Cat(), new Dog()};
for (Animal animal : animals) {
    if (animal instanceof Dog) {
        ((Dog) animal).bark();
    } else if (animal instanceof Cat) {
        ((Cat) animal).meow();
    }
}
```

Pattern Matching with `instanceof` (Java 16+)

java

```java
// Modern Java pattern matching
Object obj = "Hello World";

if (obj instanceof String s) {
    // s is automatically cast to String
    System.out.println("Length: " + s.length());
} else if (obj instanceof Integer i) {
    // i is automatically cast to Integer
    System.out.println("Value: " + i);
}


// Switch expressions with pattern matching (Java 17+)
String result = switch (obj) {
    case String s -> "String of length " + s.length();
    case Integer i -> "Integer with value " + i;
    case null -> "It's null";
    default -> "Unknown type";
};
```

The `.class` Literal - Class Objects

java

```java
// Getting Class objects
Class<String> stringClass = String.class;
```

175

```java
Class<Integer> intClass = Integer.class;

// Comparing classes
Object obj = "Hello";
if (obj.getClass() == String.class) {
    System.out.println("Exact String type");
}

// Reflection with class literals
try {
    Method lengthMethod = String.class.getMethod("length");
    Object result = lengthMethod.invoke("Hello");
    System.out.println("Length: " + result); // Length: 5
} catch (Exception e) {
    e.printStackTrace();
}
```

C++ Specific Operators

C++ offers low-level control with operators that directly manipulate memory and addresses.

Address-of Operator (&) - Getting Memory Addresses

cpp
```cpp
#include <iostream>
using namespace std;

int main() {
    int number = 42;
    int* pointer = &number; // & gets the address of number

    cout << "Value: " << number << endl;       // 42
    cout << "Address: " << &number << endl;     // Memory address
    cout << "Pointer: " << pointer << endl;     // Same address
    cout << "Value via pointer: " << *pointer << endl; // 42

    return 0;
}
```

Dereference Operator (*) - Accessing Pointer Values

cpp

```cpp
// Pointer operations
int value = 100;
int* ptr = &value;

cout << *ptr << endl;    // 100 (dereference the pointer)
*ptr = 200;              // Change value through pointer
cout << value << endl;   // 200 (original variable changed)

// Dynamic memory allocation
int* dynamicPtr = new int(50);
cout << *dynamicPtr << endl;   // 50
delete dynamicPtr;             // Free memory
```

Scope Resolution Operator (::) - Namespace and Class Access

cpp

```cpp
// Global and local variables
int globalVar = 100;

void someFunction() {
    int globalVar = 50;        // Local variable shadows global
    cout << globalVar << endl;    // 50 (local)
    cout << ::globalVar << endl;   // 100 (global using ::)
}

// Namespace resolution
namespace Math {
    const double PI = 3.14159;
    double square(double x) { return x * x; }
}

namespace Physics {
    const double PI = 3.14;   // Different value
}
```

```
int main() {
  cout << Math::PI << endl;    // 3.14159
  cout << Physics::PI << endl; // 3.14
  cout << Math::square(5) << endl; // 25
  return 0;
}
```

Ruby's Expressive Operators

Ruby has unique operators that support its philosophy of developer happiness.

The Spaceship Operator (<=>) - Three-Way Comparison

ruby
```
# The spaceship operator returns -1, 0, or 1
puts 5 <=> 3   # 1 (5 is greater than 3)
puts 3 <=> 5   # -1 (3 is less than 5)
puts 5 <=> 5   # 0 (5 equals 5)

# Useful for sorting
numbers = [3, 1, 4, 1, 5, 9, 2, 6]
sorted = numbers.sort { |a, b| a <=> b }
puts sorted   # [1, 1, 2, 3, 4, 5, 6, 9]

# String comparison
puts "apple" <=> "banana"  # -1 (apple comes before banana)
puts "zebra" <=> "apple"   # 1 (zebra comes after apple)
```

The Safe Navigation Operator (&.) - Nil-Safe Method Calls

ruby
```
# Traditional nil checking
user = nil
if user && user.profile && user.profile.name
  puts user.profile.name
end

# With safe navigation
```

```ruby
puts user&.profile&.name  # Returns nil if one step is nil

# Practical examples
users = [
  { name: "Alice", profile: { email: "alice@example.com" } },
  { name: "Bob" },  # No profile
  nil
]

users.each do |user|
  email = user&.profile&.email || "No email"
  name = user&.name || "Unknown"
  puts "#{name}: #{email}"
end
```

Range Operators (..) and (...)

ruby
```ruby
# Inclusive range (..)
puts (1..5).to_a    # [1, 2, 3, 4, 5]
puts ('a'..'e').to_a # ["a", "b", "c", "d", "e"]

# Exclusive range (...)
puts (1...5).to_a    # [1, 2, 3, 4]

# Practical uses
case age
when 0..12
  puts "Child"
when 13..19
  puts "Teenager"
when 20..64
  puts "Adult"
else
  puts "Senior"
end
```

```
# String slicing
text = "Hello World"
puts text[0..4]   # "Hello"
puts text[6..-1]  # "World"
```

PHP's Web-Oriented Operators

PHP has operators specifically designed for web development tasks.

The Null Coalescing Operator (??)

php

```php
// Default values for potentially null variables
$username = $_GET['username'] ?? 'guest';
$page = $_GET['page'] ?? 1;
$theme = $_SESSION['theme'] ?? 'light';

// Chaining null coalescing
$config = $userConfig ?? $defaultConfig ?? $systemConfig;

// Before null coalescing (verbose)
$username = isset($_GET['username']) ? $_GET['username'] : 'guest';
```

The Null Coalescing Assignment Operator (??=)

php

```php
// Assign only if variable is null
$config['theme'] ??= 'dark';      // Only set if not already set
$user['role'] ??= 'subscriber';   // Default role if none specified

// Session management
session_start();
$_SESSION['cart'] ??= [];          // Initialize cart if not exists
$_SESSION['visit_count'] ??= 0;    // Initialize counter
$_SESSION['visit_count']++;        // Increment visits
```

The Concatenation Assignment Operator (.=)

php
```
```

```php
// Building strings incrementally
$html = '<div>';
$html .= '<h1>Welcome</h1>';
$html .= '<p>This is the content</p>';
$html .= '</div>';

// Log file building
$log = date('Y-m-d H:i:s') . ' - ';
$log .= 'User logged in: ';
$log .= $_SESSION['username'];
$log .= "\n";
file_put_contents('access.log', $log, FILE_APPEND);
```

Go's Unique Operators

Go has some distinctive operators that support its simplicity and concurrency features.

The Channel Operator (<-) - Communication

```go
// Creating and using channels
ch := make(chan int)

// Sending to channel (in a goroutine)
go func() {
    ch <- 42  // Send 42 to channel
}()

// Receiving from channel
value := <-ch  // Receive value from channel
fmt.Println(value)  // 42

// Channel direction in function signatures
func sender(ch chan<- int) {  // Send-only channel
    ch <- 100
```

```go
}

func receiver(ch <-chan int) { // Receive-only channel
    value := <-ch
    fmt.Println(value)
}
```

The Short Variable Declaration (:=)

```go
// Short declaration (type inferred)
name := "Alice"         // var name string = "Alice"
age := 30               // var age int = 30
isActive := true        // var isActive bool = true

// Multiple assignment
x, y := 10, 20
name, err := getUsername()

// Only works for new variables (or at least one new variable)
var existing int
existing, new := 5, 10   // OK - 'new' is new variable
```

Practical Exercises

Exercise 1: Python Membership and Identity

```python
# Practice with Python's 'in' and 'is' operators
fruits = ['apple', 'banana', 'orange']
colors = ['red', 'yellow', 'orange']

# Your tasks:
# 1. Check if 'apple' is in fruits
# 2. Check if 'orange' is in both fruits and colors
# 3. Find items that are in both lists
# 4. Use 'is' to check if a variable is None
```

```python
def find_common_items(list1, list2):
    common = []
    for item in list1:
        if item in list2:
            common.append(item)
    return common

# Solutions:

apple_available = 'apple' in fruits
orange_in_both = 'orange' in fruits and 'orange' in colors
common_items = find_common_items(fruits, colors)

test_var = None
is_none = test_var is None

print(f"Apple available: {apple_available}")
print(f"Orange in both: {orange_in_both}")
print(f"Common items: {common_items}")
print(f"Test var is None: {is_none}")
```

Exercise 2: JavaScript Type Checking

javascript

```javascript
// Practice with typeof and instanceof
let testValues = [
    42,
    "hello",
    true,
    [],
    {},
    new Date(),
    null,
    undefined,
    function() { return "test"; }
];

// Your task: Categorize each value by type
```

```javascript
function categorizeValue(value) {
  if (typeof value === "number") {
    return "Number";
  } else if (typeof value === "string") {
    return "String";
  } else if (typeof value === "boolean") {
    return "Boolean";
  } else if (value === null) {
    return "Null";
  } else if (typeof value === "undefined") {
    return "Undefined";
  } else if (typeof value === "function") {
    return "Function";
  } else if (value instanceof Array) {
    return "Array";
  } else if (value instanceof Date) {
    return "Date";
  } else if (typeof value === "object") {
    return "Object";
  } else {
    return "Unknown";
  }
}

// Test the function
testValues.forEach((value, index) => {
  console.log(`Value ${index}: ${categorizeValue(value)}`);
});
```

Exercise 3: Multi-Language Comparison

python

```python
# Python version - membership testing
def validate_user_data(data):
    required_fields = ['name', 'email', 'age']

    # Check if all required fields are present
```

184

```python
    missing_fields = []
    for field in required_fields:
        if field not in data:
            missing_fields.append(field)

    if missing_fields:
        return False, f"Missing fields: {', '.join(missing_fields)}"

    # Validate email format
    if '@' not in data['email'] or '.' not in data['email']:
        return False, "Invalid email format"

    # Validate age
    if data['age'] is None or data['age'] < 0:
        return False, "Invalid age"

    return True, "Valid user data"

# Test cases
test_users = [
    {'name': 'Alice', 'email': 'alice@example.com', 'age': 30},
    {'name': 'Bob', 'age': 25},  # Missing email
    {'name': 'Charlie', 'email': 'invalid-email', 'age': 35},  # Invalid email
    {'name': 'Diana', 'email': 'diana@example.com', 'age': None}  # Invalid age
]

for i, user in enumerate(test_users):
    valid, message = validate_user_data(user)
    print(f"User {i + 1}: {message}")
```

Language Philosophy Through Operators

Each language's unique operators reflect its design philosophy:

Python - Readability and explicitness

- Uses English words (`in`, `is`, `not in`, `is not`)
- Emphasizes clarity over brevity
- Walrus operator added reluctantly after much debate

185

JavaScript - Flexibility and web compatibility

- `typeof` handles JavaScript's flexible typing
- `instanceof` works with prototypal inheritance
- Operators evolved to handle browser compatibility issues

Java - Type safety and object-orientation

- `instanceof` enforces type hierarchies
- `.class` supports reflection and runtime type information
- Pattern matching reduces casting boilerplate

C++ - Performance and low-level control

- Address (`&`) and dereference (`*`) operators for memory management
- Scope resolution (`::`) for namespace organization
- Direct memory manipulation capabilities

Ruby - Developer happiness and expressiveness

- Spaceship operator (`<=>`) makes sorting elegant
- Safe navigation (`&.`) prevents nil errors gracefully
- Range operators make iterations intuitive

Key Takeaways

Language-specific operators give each programming language its unique personality:

1. **Python's `in` and `is`** make membership and identity testing readable
2. **JavaScript's `typeof` and `instanceof`** handle flexible typing safely
3. **Java's `instanceof`** enforces strict type hierarchies
4. **C++'s `&` and `*`** provide direct memory manipulation
5. **Ruby's `<=>` and `&.`** prioritize developer ergonomics
6. **PHP's `??` and `.=`** support web development patterns
7. **Go's `<-` and `:=`** enable concurrency and simplicity

Understanding these differences helps you:

- Choose the right language for your project
- Write idiomatic code in each language
- Appreciate why languages evolved different solutions
- Become a more versatile programmer

These operators aren't just syntax differences—they represent different approaches to solving programming problems. Master them, and you'll think more flexibly about code.

Chapter 10: Complete Reference & Troubleshooting

Your Guide to Mastering Programming Symbols

Congratulations! You've journeyed through the intricate world of programming symbols, from the humble parentheses () to the enigmatic bitwise operators like ^ and <<. This final chapter is your one-stop reference for all the symbols we've covered, plus a troubleshooting guide to help you navigate the inevitable bumps in your coding journey. Think of this as your programming symbol survival kit—a place to revisit when you're stuck, confused, or just need a quick reminder. My goal is to ensure you feel confident using these symbols across any programming language, with practical tips to fix common errors drawn from my own struggles as a beginner.

Why a Reference and Troubleshooting Guide?

When I was learning to code, I wished for a single place where all the symbols were explained clearly, with examples and fixes for the mistakes I kept making. Tutorials often scattered this information across lessons or assumed I already knew what to do when my code threw errors like "unexpected token }" or "invalid syntax near !=." This chapter pulls everything together, organizing symbols by their purpose and pairing them with real-world troubleshooting advice. Whether you're writing your first program or debugging a stubborn bug, this guide has your back.

Comprehensive Symbol Reference

Below is a complete reference of the programming symbols covered in this book, grouped by category, with their meanings, examples in multiple languages (Python, JavaScript, Java), and common uses. Each entry is designed to be concise yet informative, so you can quickly find what you need.

Grouping Symbols

These symbols organize your code, acting like containers for logic and data.

- **Parentheses ()**
    - **Purpose**: Group expressions, define function calls, or set precedence in calculations.
    - **Example (Python)**:
    - ```
      result = (2 + 3) * 4  # Parentheses ensure 2 + 3 is calculated first
      print(result)  # Outputs: 20
      ```

    - **Example (JavaScript)**:
    - ```
      function greet(name) {
      ```

187

- o   `return `Hello, ${name}`;`
- o   `}`
  `console.log(greet("Alice"));  // Outputs: Hello, Alice`

- o **Common Use**: Ensuring order of operations, passing arguments to functions.
- **Square Brackets []**
  - o **Purpose**: Access array/list elements or define arrays.
  - o **Example (Java)**:
  - o `int[] numbers = {1, 2, 3};`
    `System.out.println(numbers[0]);  // Outputs: 1`

  - o **Common Use**: Storing and retrieving list or array data.
- **Curly Braces {}**
  - o **Purpose**: Define code blocks, objects, or dictionaries.
  - o **Example (JavaScript)**:
  - o `let user = { name: "Bob", age: 25 };`
    `console.log(user.name);  // Outputs: Bob`

  - o **Common Use**: Grouping statements in loops/functions or defining key-value pairs.

Comparison Symbols

These symbols make decisions by comparing values.

- **Equal to ==**
  - o **Purpose**: Checks if two values are equal.
  - o **Example (Python)**:
  - o `if 5 == "5":  # Type coercion in some languages`
    `    print("Equal")`

  - o **Common Use**: Testing conditions in if statements.
- **Not Equal !=**
  - o **Purpose**: Checks if two values are not equal.
  - o **Example (Java)**:
  - o `if (score != 100) {`
  - o `    System.out.println("Try again!");`
    `}`

- **Greater/Less Than >, <, >=, <=**
  - o **Purpose**: Compare numerical values.
  - o **Example (JavaScript)**:
  - o `let age = 18;`
  - o `if (age >= 18) {`
  - o `    console.log("Adult");`
    `}`

Math Symbols

These handle calculations and numerical operations.

- **Basic Operations +, -, *, /**
  - **Purpose**: Perform addition, subtraction, multiplication, division.
  - **Example (Python)**:

    ```
    total = 10 * 2 / 5  # Outputs: 4.0
    ```

- **Modulus %, Exponentiation **, Floor Division //**
  - **Purpose**: Find remainders, raise to powers, or divide with integer results.
  - **Example (Python)**:
  - ```
    print(10 % 3)   # Outputs: 1
    ```
  - ```
    print(2 ** 3)   # Outputs: 8
    print(10 // 3)  # Outputs: 3
    ```

- **Increment/Decrement ++, --, Compound Assignments +=, -=, *=**
  - **Purpose**: Modify values efficiently.
  - **Example (Java)**:
  - ```
    int x = 5;
    x += 3;   // x is now 8
    ```

Logic Symbols

These connect conditions to build complex logic.

- **Logical AND &&, OR ||, NOT !**
  - **Purpose**: Combine or invert conditions.
  - **Example (JavaScript)**:
  - ```
    if (age >= 18 && hasLicense) {
    ```
  - ```
        console.log("Can drive");
    }
    ```

- **Bitwise AND &, OR |, XOR ^, NOT ~**
  - **Purpose**: Manipulate bits directly.
  - **Example (Python)**:
  - ```
    permissions = 5 & 3  # Binary: 0101 & 0011 = 0001
    print(permissions)   # Outputs: 1
    ```

- **Shift Operators <<, >>**
  - **Purpose**: Shift bits left or right for fast multiplication/division.
  - **Example (Java)**:

    ```
    int x = 8 >> 1;   // Binary: 1000 >> 1 = 0100 (4 in decimal)
    ```

Assignment Symbols

These store or update values.

- **Basic Assignment =**
  - o **Purpose**: Assign a value to a variable.
  - o **Example (Python)**:

    ```
    x = 10
    ```

- **Compound Assignments +=, -=, *=**
  - o **Purpose**: Combine operations with assignment.
  - o **Example (JavaScript)**:
  - o ```
    let score = 100;
    score *= 2;   // score is now 200
    ```

## Special Symbols

These handle unique tasks like accessing objects or ending statements.

- **Comments //, #, /\* \*/**
  - o **Purpose**: Add notes or disable code.
  - o **Example (Java)**:
  - o `// This is a single-line comment`
  - o ```
    /* This is
       a multi-line comment */
    ```

- **Object Access ., ->**
  - o **Purpose**: Access properties or methods.
  - o **Example (JavaScript)**:
  - o ```
    let user = { name: "Alice" };
    console.log(user.name);   // Outputs: Alice
    ```

- **Statement Terminator ;**
  - o **Purpose**: End statements in languages like Java and JavaScript.
  - o **Example (Java)**:

    ```
    System.out.println("Hello");   // Semicolon required
    ```

Troubleshooting Common Symbol Errors

Even with a solid understanding of symbols, errors happen. Below are the most common issues beginners face, based on my own coding mishaps, along with how to fix them.

1. Mismatched Grouping Symbols

Error: "SyntaxError: unexpected EOF" or "missing }" **Cause**: Unclosed parentheses, brackets, or braces. **Example (JavaScript)**:

```
function example() {
    if (true) {
        console.log("Hi");
```
```

```
    // Missing closing }
```

Fix: Count your opening and closing symbols. Use an editor like Visual Studio Code that highlights matches.

```
function example() {
    if (true) {
        console.log("Hi");
    } // Added closing }
}
```

2. Confusing = with ==

Error: Code runs but produces wrong results. **Cause**: Using assignment (=) instead of comparison (==) in conditions. **Example (Python)**:

```
if x = 5:  # Assigns 5 to x, causing an error
    print("This won't work")
```

Fix: Use == for comparisons.

```
if x == 5:
    print("This works")
```

3. Mixing Logical and Bitwise Operators

Error: Unexpected logic in conditions. **Cause**: Using & instead of && or | instead of ||. **Example (Java)**:

```
if (age > 18 & hasLicense) {  // Bitwise, not logical
    System.out.println("Wrong operator");
}
```

Fix: Use logical operators for conditions.

```
if (age > 18 && hasLicense) {
    System.out.println("Correct");
}
```

4. Forgetting Semicolons in Java/JavaScript

Error: "SyntaxError: missing ; before statement" **Cause**: Omitting ; in languages that require it. **Example (Java)**:

```
int x = 5  // Missing semicolon
System.out.println(x);
```

Fix: Add semicolons where needed.

```
int x = 5;
System.out.println(x);
```

5. Incorrect Bitwise Operations

Error: Wrong numerical results. **Cause**: Misunderstanding bitwise operators like ^ or <<.
Example (Python):

```
x = 5 ^ 2   # Expected toggle, but unclear binary result
print(x)  # Outputs: 7 (binary 0101 ^ 0010 = 0111)
```

Fix: Visualize binary:

```
5: 0101
2: 0010
^: 0111 (7)
```

Test with small numbers and print intermediate results.

6. Language-Specific Symbol Differences

Error: Code works in one language but fails in another. **Cause**: Symbols like ** (Python) vs Math.pow (Java) differ. **Example**:

```
x = 2 ** 3   # Works in Python
int x = 2 ** 3;  // Fails in Java
```

Fix: Use language-appropriate symbols.

```
int x = (int) Math.pow(2, 3);  // Correct for Java
```

Visual Cheat Sheet

To make symbols stick, here's a quick visual reference:

Category	Symbols	Example (Python)
Grouping	(), [], {}	x = [1, 2, 3]
Comparison	==, !=, >, <	if x == 5:
Math	+, -, *, /, %	x = 10 % 3
Logic	&&, `	
Assignment	=, +=, -=	x += 5
Special	., ;, #, //	print(x.y)

Practice Opportunities

Reinforce your skills with these exercises:

1. **Symbol Hunt**: Write a program in Python that uses at least one symbol from each category (grouping, comparison, math, logic, assignment, special). Print a message based on a condition.
2. **Debug Challenge**: Take this buggy code and fix it:

```
3. let x = 10
4. if (x = 5 {
5.     console.log("This won't work")
   }
```

6. **Bitwise Fix**: Correct this code to check if 4 is set in permissions = 7:

```
7. permissions = 7
8. if permissions && 4:
       print("Write permission")
```

Test these in Replit, CodePen, or Visual Studio Code.

Final Tips for Success

- **Use Visual Editors**: Tools like Visual Studio Code highlight matching brackets and flag missing semicolons.
- **Test Incrementally**: Write small chunks of code and test often to catch symbol errors early.
- **Google Errors**: Search error messages with the language name (e.g., "JavaScript unexpected token }") for specific fixes.
- **Practice Daily**: Spend 5 minutes writing code with different symbols to build muscle memory.

Conclusion

You've now unlocked the hidden language of programming symbols! From organizing code with () and {} to making decisions with == and &&, you have the tools to read, write, and debug code confidently. This reference and troubleshooting guide is your companion for the journey ahead. Keep experimenting, stay curious, and don't be afraid to make mistakes—every error is a step toward mastery. Happy coding!

Glossary: Key Terms for Programming Symbols

LETTER A

Addition (+)

Definition:
The **addition operator (+)** performs **mathematical addition** or combines two values. In most programming languages, it's also used to **join (concatenate)** strings or lists.

Examples:

```
result = 5 + 3          # Outputs 8
text = "Hello" + " " + "World"  # Outputs "Hello World"
```

Used in: Math operations, string concatenation, and data merging.

Tip: In Python and JavaScript, + can work with both numbers and text, but mixing types (like adding a number to a string) may cause an error.

Angle Brackets (< >)

Definition:
Angle brackets are used to enclose tags in markup languages such as **HTML** and **XML**. They define the **start** and **end** of an element, helping structure web content.

Example:

```
<h1>Welcome to My Website</h1>
```

Used in: HTML, XML, JSX (React).

Tip: Each opening tag `<h1>` should always have a matching closing tag `</h1>` to avoid rendering errors.

Array [ ]

Definition:
An **array** is an ordered collection that stores multiple values inside **square brackets**. Each value (called an *element*) is accessed by its **index number**.

Example:

```
let colors = ["red", "green", "blue"];
console.log(colors[0]);  // Outputs: red
```

Used in: Python, JavaScript, C, Java, and most languages.

Tip: Arrays start counting from **index 0**, not 1 — a key concept called **zero-based indexing**.

Arrow Function (=>)

Definition:
An **arrow function** is a **short and modern way** to write functions in JavaScript. It's often used for cleaner, simpler code — especially in callbacks and one-line functions.

Example:

```
const add = (a, b) => a + b;
console.log(add(2, 3)); // Outputs: 5
```

Used in: JavaScript (ES6 and later).

Tip: Arrow functions don't have their own `this` keyword — which makes them ideal for simple operations but not for object methods.

Assignment (=)

Definition:
The **assignment operator (=)** is used to **store a value** inside a variable. It acts as a bridge that connects data to a variable name.

Example:

```
x = 10
```

```
message = "Hello!"
```

Used in: All programming languages.

Tip: Be careful not to confuse = (assignment) with == (comparison).
= gives a value; == checks a value.

Attribute

Definition:
An **attribute** provides **additional information** about an HTML element. Attributes appear inside a tag and are written as **name-value pairs**.

Example:

```
<img src="cat.jpg" alt="Cute cat">
```

Used in: HTML, XML, JSX (React).

Tip: Attributes define properties such as image sources, colors, IDs, or classes — making web pages more dynamic and descriptive.

LETTER B

Backslash (\)

Definition:
The **backslash (\\)** is used to **escape special characters**, meaning it tells the computer to treat the next symbol as ordinary text. It's also used for **file paths** and **newline commands**.

Examples:

```
print("Hello\nWorld")    # Outputs:
# Hello
# World
path = "C:\\Users\\Documents"
```

Used in: Python, C, C++, Java, JavaScript, and file systems.

Tip:

- \n → New line
- \t → Tab
- \\ → Literal backslash

Bitwise AND (&)

Definition:
The **bitwise AND (&)** compares each bit of two binary numbers. It outputs **1 only if both bits are 1**.
This operator is used in **low-level programming** and **hardware control**.

Example:

```
a = 5  # (0101)
b = 3  # (0011)
print(a & b)  # Outputs: 1 (0001)
```

Used in: C, C++, Python, and embedded systems.

Tip: Bitwise operations work on **binary representations** — they're essential for optimization, encryption, and device programming.

Bitwise NOT (~)

Definition:
The **bitwise NOT (~)** inverts all bits in a number — turning 1s into 0s and 0s into 1s.
It's commonly used for **complementing binary values** and performing **negative conversions**.

Example:

```
x = 5   # Binary: 0101
print(~x)  # Outputs: -6
```

Used in: C, C++, Java, Python.

Tip: The result looks unusual because computers use **two's complement** to store negative numbers.

Bitwise OR (|)

Definition:
The **bitwise OR (|)** compares bits of two numbers and outputs **1 if at least one bit is 1**.
It's useful for **combining binary flags** or **enabling options**.

Example:

```
a = 5  # (0101)
b = 3  # (0011)
print(a | b)  # Outputs: 7 (0111)
```

197

Used in: Low-level operations, device drivers, and graphics programming.

Tip: Bitwise OR is different from logical OR (| |) — one works on bits, the other on conditions.

Bitwise XOR (^)

Definition:
The **bitwise XOR (^)** — or "exclusive OR" — outputs **1 only if the bits are different**. It's a clever way to **toggle values**, **swap numbers**, or **compare differences** in binary.

Example:

```
a = 5  # (0101)
b = 3  # (0011)
print(a ^ b)  # Outputs: 6 (0110)
```

Used in: Cryptography, data encoding, and optimization.

Tip: XOR is also used in **encryption algorithms** because reversing it with the same key gives back the original data.

Boolean

Definition:
A **Boolean** represents a **true or false** value — the foundation of logic in programming. It's used to make **decisions**, **run conditions**, and **control program flow**.

Example:

```
is_logged_in = True
has_permission = False
```

Used in: All programming languages.

Tip: Booleans come from **Boolean algebra**, a 19th-century mathematical logic system that modern computers rely on.

Braces { }

Definition:
Braces — also known as **curly brackets** — group statements, code blocks, or objects together.
They define where a structure **starts and ends**.

Example:

```
if (score > 90) {
  console.log("Excellent!");
}
```

Used in: JavaScript, Java, C, C++, JSON.

Tip: Braces are essential in languages that use **block syntax** — missing one can crash your code!

Break Statement

Definition:
The **break statement** instantly **stops a loop or switch statement** when a certain condition is met.
It helps exit repetitive processes once the goal is achieved.

Example:

```
for i in range(10):
    if i == 5:
        break
print("Loop stopped at 5")
```

Used in: Python, Java, C, JavaScript.

Tip: Combine `break` with `if` for precise control of loop behavior.

Bug

Definition:
A **bug** is an **error or flaw** in code that causes unexpected results or crashes.
Bugs can range from simple typos to complex logic errors.

Example:

```
# Bug: using = instead of ==
if x = 5:   # Error
    print("Hello")
```

Used in: All programming.

Tip: The process of finding and fixing bugs is called **debugging** — and it's one of the most valuable skills a programmer can develop.

LETTER C

Compound Assignment (+=)

Definition:
The **addition compound assignment operator (+=)** adds a value to a variable and **updates** the variable in one step.
It's a shorthand for writing repetitive addition expressions.

Example:

```
x = 5
x += 3  # Same as x = x + 3
print(x)  # Outputs: 8
```

Used in: Python, JavaScript, Java, C, C++.

Tip: Compound assignment makes code cleaner and faster to write, especially inside loops or counters.

Compound Assignment (-=)

Definition:
The **subtraction compound assignment operator (-=)** subtracts a value from a variable and immediately updates it.
It simplifies repetitive subtraction operations.

Example:

```
score = 100
score -= 10  # Same as score = score - 10
print(score)  # Outputs: 90
```

Used in: Python, C, Java, JavaScript.

Tip: Great for decreasing totals, countdowns, or iterative adjustments inside loops.

Compound Assignment (*=)

Definition:
The **multiplication compound assignment operator (*=)** multiplies a variable by a value and updates it in one line.
It's often used in scaling operations or repetitive multiplication.

Example:

```
x = 4
x *= 2  # Same as x = x * 2
print(x)  # Outputs: 8
```

Used in: Python, Java, JavaScript, C, C++.

Tip: Perfect for exponential growth or scaling algorithms, such as adjusting size or speed values.

Curly Braces { }

Definition:
Curly braces — also called **braces** — enclose **code blocks, functions,** or **object definitions**. They define where a specific block of code starts and ends.

Examples:

```
if (x > 0) {
  console.log("Positive");
}
{"name": "Raquel", "age": 7}
```

Used in: C, C++, Java, JavaScript, JSON.

Tip: Every opening { must have a closing }. Mismatched braces often cause syntax errors.

LETTER D

Decrement (--)

Definition:
The **decrement operator (--)** subtracts **1** from a variable, making repetitive reductions faster. Like increment (++), it comes in two forms:

- **Pre-decrement (--x):** subtracts before using the value.
- **Post-decrement (x--):** subtracts after using the value.

LETTER E

Equal To (==)

- **Definition:**
  The **equality operator (==)** checks whether two values are **the same**.
  It returns **True** if both sides are equal and **False** otherwise.

Example:

```
int i = 10;
i--;  // Now i = 9
```

Used in: C, C++, Java, JavaScript (not in Python).

Tip: In Python, use x -= 1 since -- is not supported.

Division (/)

Definition:
The **division operator (/)** divides one number by another, returning a **decimal (floating-point)** result in most languages.

Example:

```
result = 10 / 3
print(result)  # Outputs: 3.3333333333
```

Used in: All programming languages.

Tip: To get whole numbers only, use **floor division (//)** or **integer division (div)** depending on the language.

Example:

```
x = 10
print(x == 10)  # True
```

Used in: All programming languages.

Tip:

- == compares values.
- = assigns values.
  Mixing them up is one of the most common beginner errors.

Exponentiation ()**

Definition:
The **exponentiation operator (**)** raises a number to a specified power — a shortcut for repeated multiplication.

Example:

```
print(2 ** 3)  # Outputs: 8 (2 × 2 × 2)
```

Used in: Python, JavaScript (ES6+).

Tip:
In some languages (like C or older JavaScript), use a math function instead:

```
Math.pow(2, 3); // 8
```

Floor Division (//)

Definition:
The **floor division operator (//)** divides one number by another and **rounds the result down** to the nearest whole integer.

Example:

```
print(10 // 3)  # Outputs: 3
```

Used in: Python, and some other languages as integer division.

Tip: Ideal for getting whole-number results without fractions — often used in counters, indexes, and financial calculations.

LETTER G

Greater Than (>)

Definition:
The **greater-than operator (>)** compares two values to check if the left one is **larger** than the right.
It returns **True** if the comparison is correct, otherwise **False**.

Example:

```
score = 90
if score > 80:
    print("Excellent")
```

Used in: All programming languages.

Tip: Common in sorting, filtering, and decision-making operations.

Greater Than or Equal To (>=)

Definition:
The **greater-than-or-equal-to operator (>=)** checks whether one value is **larger than or equal to** another.
It's useful when you need to include the upper boundary in your condition.

Example:

```
temperature = 32
if temperature >= 32:
```

```
print("Water can freeze")
```

Used in: All programming languages.

Tip: Always use >= for **inclusive comparisons**, especially in ranges or loop conditions.

LETTER H

Hash (#)

Definition:
A **hash** symbol (#) is used in programming to mark a **comment** or note that the computer ignores. It helps programmers explain what their code does.

Example (Python):

```
# This prints a welcome message
print("Hello, world!")
```

Used in: Python, Bash, YAML.
Tip: Comments make your code easier to read and maintain — always describe what tricky lines do.

Hashtag (#)

Definition:
A **hashtag** is the same symbol (#) used in **social media**, not code. When placed in front of a word, it creates a **searchable tag**.

Example:

```
#coding    #LearnPython    #WebDevelopment
```

In Coding Context:
In **Markdown files** (used for documentation), a hashtag also creates **headings**.

Example (Markdown):

```
# Title
## Subtitle
### Subsection
```

Hash Function

Definition:
A **hash function** is a mathematical process that converts input data (like a word or file) into a **fixed-length code** called a **hash value**.
It's like a unique digital fingerprint for data.

Example:

```
import hashlib
result = hashlib.sha256(b"hello").hexdigest()
print(result)
# Outputs: 2cf24dba5fb0...
```

Key Uses:

- Verifying file integrity
- Password storage and encryption
- Quick data lookup in databases

Tip:
Hash functions are **one-way** — you can create a hash, but you can't reverse it to see the original data.

Hash Table

Definition:
A **hash table** is a **data structure** that stores information in **key–value pairs**, allowing very fast lookups.
It uses a **hash function** to find where data should go.

Example (Python Dictionary):

```
students = {"Alice": 90, "Bob": 85, "Charlie": 95}
print(students["Alice"])  # Outputs: 90
```

Key Uses:

- Storing large data efficiently
- Used in dictionaries, maps, and caches

Tip:
Hash tables are one of the most common and powerful tools in programming for organizing data.

Header Tags (<h1>–<h6>)

Definition:
In **HTML**, header tags define titles or sections of a webpage. `<h1>` is the largest and most important, while `<h6>` is the smallest.

Example:

```
<h1>Main Title</h1>
<h2>Subheading</h2>
<h3>Smaller Section</h3>
```

Used in: Web development (HTML/CSS).
Tip: Always use only one `<h1>` per page — it helps with SEO and structure.

Hexadecimal (HEX)

Definition:
Hexadecimal is a base-16 number system used to represent colors, memory addresses, and binary data in a compact way.
It uses digits 0–9 and letters A–F.

Example:

```
#FF0000 = Red
#00FF00 = Green
#0000FF = Blue
```

Used in: HTML, CSS, programming memory, and debugging.
Tip: "Hex" codes are just a readable way to show long binary values.

Highlighting

Definition:
Syntax highlighting is a feature in code editors that adds color to code — keywords, symbols, and comments — to make reading and debugging easier.

Example:
In VS Code, functions might appear blue, strings green, and comments gray.

Used in: Code editors, IDEs, online coding platforms.
Tip: If your code is all black text, enable syntax highlighting in your editor!

HTML (HyperText Markup Language)

Definition:
HTML is the main language used to build websites. It uses **tags** enclosed in angle brackets (< >) to structure text, images, and links.

Example:

```
<p>Hello, world!</p>
```

Used in: Web development, front-end design.
Tip: HTML controls the content and layout, while CSS handles colors and style.

HTTP / HTTPS

Definition:
HTTP (HyperText Transfer Protocol) and **HTTPS** (its secure version) are used to transfer data between your browser and a website.

Example:

```
http://example.com
https://secure-site.com
```

Used in: Web addresses (URLs).
Tip: Always use **HTTPS** — it encrypts the connection and protects user data.

Hyphen (-)

Definition:
The **hyphen** is a short dash used for **joining words**, **separating elements**, or **representing subtraction** in code. In programming, the hyphen can serve multiple purposes depending on where it appears:

- As a **mathematical operator**, it performs **subtraction** between two values.
- As a **negative sign**, it indicates a value less than zero.
- In **file names**, it separates words for readability.
- In **CSS**, it connects multi-word property names.

Example:

```
result = 10 - 3   # subtraction
balance = -5      # negative number
```

Used in: Math operations, variable naming, URLs, and CSS properties.

Tip: In CSS, hyphens are common in property names:

```
font-size: 16px;
background-color: red;
```

Increment (++)

Definition:

The **increment operator (++)** increases the value of a variable by **one**. It's a shortcut that makes counters and loops concise and efficient.

There are two forms:

- **Pre-increment (++x)** → increases the value *before* it's used.
- **Post-increment (x++)** → increases the value *after* it's used.

Example:

```
int count = 5;
count++;    // Now count = 6
```

Used in: C, C++, Java, JavaScript (not in Python).

Tip: Python doesn't use ++; instead, write `count += 1`.

LETTER L

Less Than (<)

Definition:

The **less-than symbol (<)** compares two values to see if one is **smaller** than the other.
It returns **True** if the first value is smaller and **False** otherwise.

Example:

```
x = 5
print(x < 10)   # True
```

Used in: Conditional statements and loops.

Tip: It's often used in loops that run while a condition remains true:

```
while x < 10:
    x += 1
```

Less Than or Equal To (<=)

Definition:

The **less-than-or-equal-to symbol (<=)** tests whether one value is **smaller than or equal to** another.

It's ideal for creating **inclusive limits** in comparisons or loops.

Example:

```
age = 17
if age <= 18:
    print("Eligible for student discount")
```

Used in: Comparisons, loop boundaries, and conditional logic.

Tip: Use `<=` to include the boundary value — for example, counting up to 10 includes 10 itself.

Logical AND (&&)

Definition:

The **logical AND operator (&&)** checks whether **two or more conditions** are true at the same time.

It returns **True** only if **all** conditions are true.

Example:

```
if (age > 18 && age < 65) {
  console.log("Working age");
}
```

Used in: Conditional statements, boolean expressions, and flow control.

Tip: In Python, `and` replaces `&&`:

```
if age > 18 and age < 65:
    print("Working age")
```

Logical NOT (!)

Definition:

The **logical NOT operator (!)** flips or reverses a condition's truth value.

If a condition is **True**, `!` makes it **False**, and vice versa.

Example:

```
let loggedIn = false;
if (!loggedIn) {
  console.log("Please log in");
```

```
}
```

Used in: Boolean logic, negation, and conditional expressions.

Tip: In Python, `not` replaces `!`:

```
if not logged_in:
    print("Please log in")
```

Logical OR (||)

Definition:
The **logical OR operator** (`||`) checks whether **at least one** of two conditions is true.
If any one of them is true, the result is **True**.

Example:

```
if (day === "Saturday" || day === "Sunday") {
  console.log("Weekend!");
}
```

Used in: Decision-making, conditions, and branching.

Tip: In Python, `or` replaces `||`:

```
if day == "Saturday" or day == "Sunday":
    print("Weekend!")
```

LETTER M

Modulus (%)

Definition:
The **modulus operator** (`%`) finds the **remainder** when one number is divided by another.
It's commonly used for **checking divisibility**, **rotating sequences**, or **creating patterns** in loops.

Example:

```
print(10 % 3)   # Outputs: 1
```

Used in: Math, loops, algorithms (like checking even or odd numbers).

Tip:
To check if a number is even:

```
if x % 2 == 0:
    print("Even")
```

Multiplication (*)

Definition:
The **asterisk (*)** is used for **multiplication** in mathematics and programming.
It can also serve special meanings such as:

- **Repetition** of strings or lists.
- **Unpacking** values in Python.

Example:

```
result = 4 * 3        # 12
print("Hi" * 3)       # HiHiHi
numbers = [1, 2, 3]
print(*numbers)       # Unpacks list → 1 2 3
```

Used in: Math, string repetition, and data unpacking.

Tip: The * symbol has many advanced uses in programming, such as variable arguments
(*args).

Not Equal (!=)

Definition:
The **not equal operator (!=)** checks whether two values are **different**.
It returns **True** if the values are unequal, and **False** if they match.

Example:

```
password = "1234"
if password != "0000":
    print("Access granted")
```

Used in: Comparisons, conditionals, and loops.

Tip: In Python and most languages, != means "not equal." In older systems like BASIC, <>
was used instead.

Parentheses ( )

Definition:
Parentheses (also called **round brackets**) group expressions, define function calls, and

control the **order of operations** in code.
They tell the computer what to evaluate first.

Examples:

```
result = (10 + 5) * 2  # Calculates 15 first, then multiplies
print("Hello, world!")  # Function call
```

Used in: Math operations, function calls, and logical grouping.

Tip: Parentheses clarify meaning — use them to avoid errors caused by operator precedence.

LETTER S

Scope (:: or {})

Definition:
Scope defines **where** in your program a variable or function can be accessed.
Symbols like {} (curly braces) or :: (scope resolution operator in C++) indicate these boundaries.

Example (C++):

```
int x = 5;
{
    int y = 10; // y only exists here
}
```

Used in: C, C++, Java, JavaScript.
Tip: Think of scope like "permission zones" for variables.

Semicolon (;)

Definition:
A **semicolon** is a punctuation symbol used to **separate statements** in many programming languages like Java, JavaScript, and C++.
It tells the computer that one instruction has ended and another is beginning.

Example:

```
let name = "Susie";
console.log(name);
```

Used in: Java, JavaScript, C, C++, PHP.
Tip: Python does not require semicolons — it uses new lines instead.

Separator (, or ;)

Definition:
A **separator** is a symbol that divides items, arguments, or statements.
The most common separators are commas (,) and semicolons (;).

Example:

```
print("A", "B", "C")  # Comma separates values
```

Used in: Python, Java, JavaScript, C++.

Single Quote (')

Definition:
A **single quote** defines a **string** — a sequence of text — in many programming languages.

Example:

```
name = 'Raquel'
```

Used in: Python, JavaScript, PHP.
Tip: You can use single or double quotes for strings, but be consistent throughout your code.

Slash (/) and Backslash (\ )

Definition:
The **slash (/)** and **backslash (\ )** look similar but serve different purposes:

- **Slash (/)** divides or separates (used in file paths or division).
- **Backslash (\ )** is used to **escape** special characters in strings.

Examples:

```
path = "C:\\Users\\Documents"
result = 10 / 2  # division
```

Used in: File systems, math operations, and escape sequences.

Square Brackets [ ]

Definition:
Square brackets are used to define **lists or arrays** — collections of items stored together.

Example:

```
numbers = [1, 2, 3, 4, 5]
```

```
print(numbers[0])  # Outputs: 1
```

Used in: Python, JavaScript, JSON.
Tip: Always remember — indexing starts at **0**, not 1!

String (" " or ' ')

Definition:
A **string** is text inside quotes. It can include words, symbols, or even numbers.

Example:

```
greeting = "Hello, world!"
```

Used in: All major programming languages.
Tip: Strings are for words; numbers are for math.

Syntax

Definition:
Syntax refers to the **rules of writing code** — like grammar in English.
It defines how symbols, punctuation, and keywords must be arranged so the computer understands your instructions.

Example:
☑ Correct Syntax

```
print("Hello, world!")
```

✖ Incorrect Syntax

```
print("Hello, world!"
```

Used in: Every programming language.
Tip: A **syntax error** happens when something is written incorrectly, such as a missing parenthesis or colon.

Syntax Error

Definition:
A **syntax error** occurs when your code breaks the language's grammar rules, preventing it from running.

Example:

```
if x = 5  # Missing colon
    print(x)
```

Used in: All programming languages.
Tip: Read the error message — it usually tells you **where** the mistake is and what symbol is missing.

Statement

Definition:
A **statement** is a single instruction that performs an action, such as printing text or assigning a value.

Example:

```
x = 10
print(x)
```

Used in: All languages.
Tip: Every statement is a building block in your program — one clear step at a time.

String Concatenation (+)

Definition:
Concatenation means joining two or more strings together using the plus (+) symbol.

Example:

```
first = "Hello"
second = "World"
print(first + " " + second)
```

Used in: Python, JavaScript, PHP, Java.
Tip: In Python 3.6+, you can also use **f-strings** for cleaner output:

```
name = "Raquel"
print(f"Hello, {name}")
```

Semantics

Definition:
Semantics refers to the **meaning** of code — what your program actually does when executed. Syntax is about **structure**; semantics is about **function**.

Example:
Both of these lines are correct syntax, but they do different things:

```
x = 5 + 2    # Adds numbers
x = "5" + "2"  # Joins strings
```

Tip: Syntax makes code run. Semantics makes it do the **right thing**.

Symbol

Definition:
A **symbol** is any special character that performs a specific task in code — such as +, -, {}, or
==.

Examples:

- = → Assignment
- + → Addition or concatenation
- {} → Code block

Tip: Every symbol has a role — once you learn them, you'll "speak" code fluently!

Space ( )

Definition:
A **space** might look empty, but in coding, it matters!
Spaces separate variables, words, and arguments so the computer knows where one thing ends
and another begins.

Example:
`print("Hello, world!")`
✗ `print ("Hello, world!")` — adds unnecessary space.

Used in: All languages.
Tip: Some languages (like Python) use **indentation and spaces** as part of their syntax.

Slash Comment (//)

Definition:
In many languages, // starts a **single-line comment** that explains the code but isn't executed.

Example:

```
// This line prints a name
console.log("Susie");
```

Used in: Java, JavaScript, C, C++.

Scope Resolution (::)

Definition:
A **scope resolution operator** (::) allows access to identifiers or variables defined in a specific scope.

Example:

```
std::cout << "Hello";
```

Used in: C++, PHP.
Tip: It helps when two parts of code have variables with the same name but different purposes.

Separator Line (——)

Definition:
A **separator line** (like dashes or underscores) is not read by the compiler but helps visually divide sections in code or documentation.

Example:

```
# ---------- Start of Function ----------
```

Semitransparent Symbols (in UI)

Definition:
Some programming editors show **faint or semitransparent symbols** to indicate formatting marks like tabs or spaces.
They're visual guides — not part of the actual code.

Tip: These can usually be toggled on or off in your code editor settings.

Semicolon ;
   A symbol that ends statements in languages like Java and JavaScript, ensuring clear separation.
Square Brackets []
   Symbols used to define or access elements in lists, arrays, or dictionaries.
Subtraction -
   A symbol that subtracts one value from another, for numerical operations or counters.

217

LETTER T

Tab (\t)

Definition:
A **tab** is a special whitespace character used to **indent code** or create spacing between elements.
In Python and other languages, indentation (spaces or tabs) defines code blocks.

Example:

```
for name in ["Addison", "Raquel"]:
    print(name)   # This line is indented with a tab
```

Used in: Python, text editors, and IDEs.
Tip: Always use **consistent indentation** (either 4 spaces or a tab) to avoid syntax errors.

Tab Character (\t)

Definition:
A **tab character** is a hidden symbol that represents a specific number of spaces in a string or output.

Example:

```
print("Name\tAge")
print("Raquel\t7")
```

Output:

```
Name    Age
Raquel  7
```

Used in: Text formatting, console displays, file writing.
Tip: Tabs are handy for neat columns — especially in CSV or table-like data.

Ternary Operator (?:)

Definition:
A **ternary operator** lets you write a short **if–else statement** on one line.
It's a compact way to assign a value based on a condition.

Example (JavaScript):

```
let age = 18;
let status = (age >= 18) ? "Adult" : "Minor";
```

Used in: JavaScript, Java, C, C++, PHP, Python (as an inline if).
Tip: Read it as: *"if this is true, use that; otherwise, use something else."*

Tilde (~)

Definition:
The **tilde** symbol (~) performs a **bitwise NOT operation**, flipping binary digits.
In some systems, it also represents a user's **home directory**.

Examples:

```
x = 5      # Binary: 0101
print(~x) # Output: -6   (inverts bits)
cd ~    # Takes you to your home folder
```

Used in: Python, C, Bash, Linux file paths.
Tip: In file paths, ~ is shorthand for "your personal folder."

Token

Definition:
A **token** is the smallest unit of meaning in code — a keyword, symbol, or identifier.
When code runs, the compiler breaks everything into **tokens** to understand what you wrote.

Example:

```
x = 10
print(x)
```

Tokens here are: x, =, 10, print, (, x,)

Used in: All programming languages.
Tip: Tokens are like words in a sentence — the building blocks of your code.

Try / Except (or Try / Catch)

Definition:
The **try/except** (Python) or **try/catch** (Java, JavaScript) block handles **errors** gracefully, so your program doesn't crash.

Example (Python):

```
try:
    result = 10 / 0
except ZeroDivisionError:
    print("You can't divide by zero!")
```

Used in: Python, Java, JavaScript, C++.
Tip: Use it when something "might go wrong" — like opening a missing file or dividing by zero.

Template Literal (Backticks)

Definition:
A **template literal** allows you to combine variables directly into strings using backticks.

Example (JavaScript):

```
let name = "Raquel";
console.log(`Hello, ${name}!`);
```

Used in: JavaScript (ES6+).
Tip: Template literals make strings more readable and dynamic.

Text Editor

Definition:
A **text editor** is a tool used to write and edit code — examples include **VS Code**, **Sublime Text**, and **Notepad++**.

Tip: Code editors highlight syntax, auto-indent your text, and often include built-in debugging tools.

Throw Statement

Definition:
A **throw** statement signals an error or "exception" intentionally, allowing developers to handle it with `try/catch`.

Example (JavaScript):

```
throw new Error("Something went wrong!");
```

Used in: JavaScript, Java, C++.
Tip: "Throw" means "stop and handle this problem right now."

True / False

Definition:
True and **False** are Boolean values that represent logic decisions — something either *is* or *is not*.

Example:

```
is_open = True
is_closed = False
```

Used in: All programming languages.
Tip: Computers think in True/False logic — it's the foundation of decision making in code.

Try Block

Definition:
The **try block** tests a section of code for errors, and if something fails, the **except** or **catch** block handles it.

Example:

```
try:
    open("file.txt")
except:
    print("File not found!")
```

Tip: It's better to handle an error than let your entire program crash.

Tuple ( )

Definition:
A **tuple** is a **collection of items** stored in parentheses. It looks like a list but **cannot be changed** after creation.

Example:

```
colors = ("red", "green", "blue")
print(colors[0])  # Outputs: red
```

Used in: Python.
Tip: Use tuples when you want data that must stay **unchanged** — like days of the week or coordinates.

Turtle Graphics (Python)

Definition:
Turtle graphics is a simple module in Python that uses a turtle cursor to draw shapes and symbols on screen.

Example:

```
import turtle
turtle.forward(100)
```

```
turtle.done()
```

Used in: Python (beginner programming and symbol drawing).
Tip: Great for learning loops, coordinates, and geometry visually.

Type Casting

Definition:
Type casting converts one data type into another, such as turning text into a number.

Example:

```
x = "5"
y = int(x)    # Converts string to integer
print(y + 10)  # Outputs: 15
```

Used in: Python, C, JavaScript, Java.
Tip: Always check the type before conversion to avoid runtime errors.

Type Checking (typeof)

Definition:
Type checking ensures that variables are the correct kind before you use them.
The `typeof` keyword (in JavaScript) or `type()` function (in Python) reveals a variable's type.

Example:

```
typeof 42         // "number"
typeof "Susie"    // "string"
```

Used in: Python, JavaScript, TypeScript, Java.
Tip: Prevents logic bugs — like adding text and numbers by accident.

Type (Data Type)

Definition:
A **type** tells the computer what kind of data something is — number, text, list, etc.

Example:

```
x = 5         # integer
y = "Hello"   # string
z = [1, 2]    # list
```

Used in: All programming languages.
Tip: Knowing the type prevents mistakes like adding text to numbers.

TypeError

Definition:
A **TypeError** occurs when you try to perform an operation on the wrong type of data — like adding a string to a number.

Example:

```
print("5" + 5)   # TypeError
```

Used in: Python, JavaScript.
Tip: Always check variable types before combining them.

Typing System (Static vs Dynamic)

Definition:
Programming languages have different ways of handling data types:

- **Static typing:** Types are checked **before** the program runs (e.g., Java, C++).
- **Dynamic typing:** Types are checked **while** the program runs (e.g., Python, JavaScript).

Example:
In Python, you can assign any type freely:

```
x = 5
x = "Hello"   # Allowed
```

Tip: Dynamic typing is flexible but can lead to unexpected bugs. Static typing catches errors early.

LETTER U

Unary Operator (+, -)

Definition:
A **unary operator** is an operator that works on **only one operand** (one value).
Common unary operators are +, -, and !.

Example:

```
x = -5      # Negative number (unary minus)
y = +5      # Positive number (unary plus)
```

Used in: Python, Java, C++, JavaScript.
Tip: Unary operators often change a value's **sign** or **state** (e.g., from positive to negative or True to False).

Underscore (_)

Definition:
The **underscore** is a small symbol often used to **replace spaces** in variable names or to mark "throwaway" variables.

Examples:

```
user_name = "Raquel"
for _ in range(5):    # Loop runs 5 times, variable not needed
    print("Hi!")
```

Used in: Python, C, JavaScript, filenames.
Tip: Use underscores for clarity when naming variables:
☑ `first_name` instead of ✖ `firstname`.

Unicode (U+0000)

Definition:
Unicode is a universal standard that assigns a unique code number to every character and symbol in every language.

Example:

```
print("\u2764")   # Outputs: ♥
```

Used in: All programming languages.
Tip: Unicode makes it possible to use **emoji, math symbols, and global languages** in your programs.

Unindent / Indentation Error

Definition:
An **unindent error** happens when code is not aligned properly.
In languages like Python, **indentation shows structure**, so uneven spaces cause errors.

Example:
✖ Incorrect:

```
if True:
print("Hello")    # Missing indent
```

☑ Correct:

```
if True:
    print("Hello")
```

Used in: Python.
Tip: Always keep your indents consistent — 4 spaces per level is standard.

Unpacking

Definition:
Unpacking lets you assign multiple values from a list or tuple to variables in one line.

Example:

```
x, y, z = [1, 2, 3]
print(x)  # Outputs: 1
```

Used in: Python.
Tip: You can use an asterisk * to unpack extra items:

```
a, *b = [1, 2, 3, 4]
print(b)  # Outputs: [2, 3, 4]
```

Update Operator (+=, -=, =, /=)

Definition:
Update operators combine an operation and assignment in one step.
They modify a variable's value without rewriting the entire line.

Example:

```
x = 5
x += 2    # Same as x = x + 2
print(x)  # Outputs: 7
```

Used in: Python, C, C++, Java, JavaScript.
Tip: Great for counters and loops — less typing, same result.

URL (Uniform Resource Locator)

Definition:
A **URL** is the address of a webpage or online file.
It tells your browser where to find information on the internet.

Example:

```
https://www.youraistudybuddy.com
```

Used in: HTML, web development, APIs.
Tip: URLs are case-sensitive after the domain name — keep them simple and lowercase.

User Input

Definition:
User input allows programs to receive data typed by the user while running.

Example:

```
name = input("What is your name? ")
print("Hello, " + name)
```

Used in: Python, JavaScript, Java, C++.
Tip: Always **validate** input — users might enter something unexpected.

User Interface (UI)

Definition:
A **User Interface** is what users see and interact with — the buttons, menus, and screens of a program.

Example:

- A website form
- A mobile app layout
- A chatbot window

Used in: Web design, software development, app design.
Tip: A clear, simple UI helps users understand your code's purpose — design for ease, not complexity.

Utility Function

Definition:
A **utility function** is a small, reusable function that performs a common task — like formatting text or calculating totals.

Example:

```
def square(x):
    return x * x

print(square(4))  # Outputs: 16
```

Used in: All programming languages.
Tip: Create utility functions to avoid repeating code — it keeps your programs neat and modular.

Value

Definition:
A **value** is the actual piece of data stored in a variable — like a number, text, or result of a calculation.

Example:

```
x = 10        # 10 is the value
name = "Sue"  # "Sue" is the value
```

Used in: All programming languages.
Tip: Every variable has a **name** (identifier) and a **value** that can change as the program runs.

Variable (=)

Definition:
A **variable** is like a labeled container that stores data. The = symbol is the **assignment operator**, used to give a value to that container.

Example:

```
age = 7
message = "Hello, Raquel!"
```

Used in: Every programming language.
Tip: Variables make programs flexible — instead of hardcoding numbers, you can store values that can be updated anytime.

Variable Declaration

Definition:
Variable declaration means introducing a variable to your program so it can be used later. Some languages require explicit declarations; others do it automatically.

Example (Java):

```
int score = 90;
```

Used in: Java, C, C++, JavaScript.
Tip: Declaring variables keeps your code organized and prevents naming conflicts.

Variable Naming Rules

Definition:
When naming variables, you must follow certain **syntax rules** so your program runs without errors.

Rules:

- Must start with a **letter** or **underscore**
- Can include numbers but not start with one
- No spaces (use underscores instead)
- Case-sensitive (`Name ≠ name`)

Examples:

☑ `user_age, total_amount, _score`

✖ `2name, user age, total-amount`

Variable Scope

Definition:
Scope determines where a variable can be used or accessed inside a program.

Example (Python):

```
def greet():
    name = "Addison"   # local variable
    print(name)

greet()
# print(name)   # Error: name not accessible outside function
```

Used in: All programming languages.
Tip: Local variables exist inside a function; global variables exist everywhere.

Variable Type (Data Type)

Definition:
Each variable has a **data type**, which defines what kind of value it holds — numbers, strings, lists, or Booleans.

Example:

```
num = 25        # Integer
word = "Book"   # String
flag = True     # Boolean
```

Used in: All programming languages.
Tip: Choosing the correct type helps your code run smoothly and avoid type errors.

Vector [ ]

Definition:
A **vector** is an ordered list of numbers or elements stored in a one-dimensional array.
It's used in mathematics, physics, and programming for storing data like coordinates or numerical values.

Example (Python using NumPy):

```
import numpy as np
v = np.array([1, 2, 3])
print(v)
```

Used in: Python, C++, Java, data science.
Tip: Think of a vector as a **list of values in a straight line** — used in AI, graphics, and scientific computing.

Vector Graphics

Definition:
Vector graphics use mathematical formulas to represent shapes and lines — not pixels. This allows them to scale without losing quality.

Example:
SVG (Scalable Vector Graphics) files are used for icons and logos on websites.

Used in: Web design, AI image generation, game development.
Tip: Vectors are great for sharp, scalable images and symbols in digital art or coding interfaces.

Version Control (Git)

Definition:
Version control tracks changes made to code so you can view history, revert mistakes, and collaborate with others.
The most popular tool for this is **Git**.

Example (Git command):

```
git commit -m "Added new function"
```

Used in: Software development, team projects, GitHub, GitLab.
Tip: Version control is like a "time machine" for your code — you can always go back to earlier versions safely.

Virtual Environment (venv)

Definition:
A **virtual environment** is a sandbox for Python projects — it isolates your project's dependencies so they don't interfere with others.

Example (Command):

```
python -m venv myenv
```

Used in: Python programming.
Tip: Always activate your virtual environment before installing new packages — it keeps your project clean.

Void

Definition:
In some programming languages, **void** means a function **does not return a value**.
It performs an action but doesn't give back data.

Example (C++):

```
void greet() {
   cout << "Hello!";
}
```

Used in: C, C++, Java.
Tip: Think of `void` as a "do something" command — not a "give something back" function.

Volatile

Definition:
The keyword **volatile** tells the computer that a variable's value might change unexpectedly, even if it looks constant.
It's often used when hardware or other threads can modify data.

Example (C):

```
volatile int flag = 0;
```

Used in: C, C++, embedded systems.
Tip: `volatile` prevents the compiler from making unsafe assumptions about variable stability — used in low-level or sensor-related coding.

LETTER W

Walrus Operator (:=)

Definition:
The **walrus operator** (:=) assigns a value to a variable **as part of an expression**.
It allows you to create and use a variable in the same line of code.

Example (Python 3.8+):

```
if (n := len("Raquel")) > 3:
```

```
print(n)
```

Used in: Python (since version 3.8).
Tip: It's called a "walrus" because the symbol looks like eyes and tusks (:=).

Warning Message

Definition:
A **warning** indicates a potential issue in code that doesn't stop the program but suggests caution.

Example (Python):

```
DeprecationWarning: This function will be removed in future versions.
```

Used in: All programming languages.
Tip: Don't ignore warnings — they help you future-proof your code.

Web Address (URL)

Definition:
A **web address** (URL) tells browsers where to find a website or resource on the internet.

Example:

```
https://www.pinnaclefilingsystemonline.com
```

Used in: Web development, APIs, browsers.
Tip: Always begin secure sites with **https://** for encrypted connections.

Web Tag (< >)

Definition:
A **web tag** is a pair of angle brackets used in HTML to mark the start and end of an element. Everything inside defines part of a web page.

Example:

```
<p>Hello, world!</p>
```

Used in: HTML, XML.
Tip: Always close your tags properly — unclosed tags can break the structure of a webpage.

While Loop

Definition:
A **while loop** repeats code **as long as a condition is true**.

Example:

```
x = 1
while x <= 5:
    print(x)
    x += 1
```

Used in: Python, Java, C, JavaScript.
Tip: Be careful with conditions — if it never becomes false, your loop runs forever.

While True Loop

Definition:
A **while True** loop runs endlessly until a `break` statement stops it.

Example:

```
while True:
    print("Running...")
    break
```

Used in: Python, C, Java.
Tip: Always include a stopping condition — infinite loops can freeze your program.

Whitespace ( )

Definition:
Whitespace includes spaces, tabs, and new lines that separate code elements.
While it's invisible, it affects how your code runs and reads.

Example:

```
print("Hello, world!")  # Correct
print ("Hello, world!") # Extra space (still works, but messy)
```

Used in: All programming languages.
Tip: In Python, **indentation (whitespace)** defines blocks — so spacing must be exact.

Whitespace Character (\n, \t, space)

Definition:
Whitespace characters include spaces, tabs (\t), and new lines (\n).
They format text but are invisible.

Example:

```
print("Hello\nWorld")
```

Output:

```
Hello
World
```

Tip: In Python, whitespace also defines code blocks — indentation is not optional!

Widget

Definition:
A **widget** is a small, interactive component in a user interface, like a button, text box, or dropdown menu.

Example:
HTML widgets:

```
<input type="text" placeholder="Enter name">
```

Used in: Web development, GUIs, apps.
Tip: Widgets make interfaces user-friendly — they connect human actions to code functions.

Wildcard (*, ?)

Definition:
A **wildcard** symbol represents **one or more unknown characters**.
They are commonly used in file searches and regular expressions.

Examples:

```
*.txt    → matches all text files
data?.csv → matches data1.csv, data2.csv, etc.
```

Used in: File systems, command lines, search patterns.
Tip: Wildcards help find or select multiple items without typing every name.

Window Object

Definition:
In web programming, the **window** object represents the browser window — it's the global object in JavaScript.

Example:

```
window.alert("Welcome!");
```

Used in: JavaScript, web APIs.
Tip: Most browser functions (like alerts or prompts) belong to the `window` object.

Wrapper Function

Definition:
A **wrapper function** is a small function that adds extra features or modifies how another function works.

Example (Python):

```
def greet(func):
    def wrapper():
        print("Hello!")
        func()
    return wrapper
```

Used in: Python (decorators), web frameworks.
Tip: Wrappers help you reuse logic — ideal for logging, security, or formatting tasks.

Write / Read (I/O Operations)

Definition:
Write means saving data to a file.
Read means loading data from a file.
Together, they form **I/O (Input/Output) operations.**

Example (Python):

```
# Writing
with open("notes.txt", "w") as file:
    file.write("Hello")

# Reading
with open("notes.txt", "r") as file:
    print(file.read())
```

Used in: All programming languages.
Tip: Always close files or use `with` to avoid losing data.

Word Boundary (\b)

Definition:
A **word boundary** (\b) is a symbol in regular expressions that marks the start or end of a word.

Example (Regex):

```
\bcat\b
```

Matches "cat" but not "catalog."

Used in: Regular expressions (Regex).
Tip: Boundaries are essential when searching for full words only.

Workflow

Definition:
A **workflow** is the sequence of steps a program, developer, or automation follows to complete a task.
In coding, it often includes writing, testing, and deploying code.

Example:

- Write → Save → Run → Debug → Deploy

Used in: Software development, automation, DevOps.
Tip: Create a consistent workflow — it saves time and reduces mistakes.

Wrapper Class

Definition:
A **wrapper class** converts primitive data types (like `int` or `float`) into objects so they can be used with advanced features like lists or generics.

Example (Java):

```
Integer num = 10;  // int wrapped into Integer object
```

Used in: Java, C#, .NET.
Tip: Wrapper classes bridge simple data and object-oriented code.

LETTER X

X-Axis

Definition:
The **X-axis** is the **horizontal line** in graphs or coordinate systems.
In programming, it's used to represent **width**, **horizontal distance**, or **time** in visualizations.

Example (Python – Matplotlib):

```
import matplotlib.pyplot as plt
plt.plot([1, 2, 3], [4, 5, 6])
plt.xlabel("X-Axis")
plt.ylabel("Y-Axis")
plt.show()
```

Used in: Data visualization, AI graphics, game design.
Tip: Think of the X-axis as "left to right" — it controls horizontal placement.

XHTML (Extensible HyperText Markup Language)

Definition:
XHTML is a stricter version of **HTML** that follows **XML rules**.
It ensures cleaner, more consistent code for web browsers.

Example:

```
<html xmlns="http://www.w3.org/1999/xhtml">
  <body>
    <p>Hello, world!</p>
  </body>
</html>
```

Used in: Web development and legacy sites.
Tip: Every tag in XHTML must be **properly closed**, and attributes must be in **quotes**.

XML (Extensible Markup Language)

Definition:
XML is a markup language that stores and organizes data in a structured way using **custom tags**.

Example:

```
<student>
  <name>Raquel</name>
  <age>7</age>
</student>
```

Used in: Data storage, configuration files, APIs.
Tip: XML looks like HTML but is for **data**, not for displaying web pages.

XML Parser

Definition:
An **XML parser** reads XML files and converts them into a form programs can understand and manipulate.

Example (Python):

```python
import xml.etree.ElementTree as ET
tree = ET.parse('data.xml')
root = tree.getroot()
print(root.tag)
```

Used in: Data exchange, web services.
Tip: Always handle XML parsing carefully — incorrect syntax can cause program errors.

XML Schema (XSD)

Definition:
An **XML Schema Definition (XSD)** defines the **structure and rules** of an XML document — what tags can appear and what data they contain.

Example:

```xml
<xs:element name="student">
  <xs:complexType>
    <xs:sequence>
      <xs:element name="name" type="xs:string"/>
      <xs:element name="age" type="xs:integer"/>
    </xs:sequence>
  </xs:complexType>
</xs:element>
```

Used in: Data validation, system integration, APIs.
Tip: Think of XSD as a **blueprint** that keeps XML data organized and consistent.

XPath (/ /)

Definition:
XPath is a query language used to **navigate and locate nodes** in XML or HTML documents. It uses slashes (/) to move through elements.

Example:

```
/students/student/name
```

Used in: Web scraping, XML parsing, automation tools.
Tip: Think of XPath like a "map" for XML — it tells your code where specific data lives.

XOR (^)

Definition:
XOR stands for **Exclusive OR** — a logical operation that returns `True` **only if one condition is true**, but not both.

Example (Python):

```
a = True
b = False
print(a ^ b)  # Outputs: True
```

Used in: Logic, encryption, digital circuits.
Tip: XOR is useful in problems that involve **flipping values** or **checking differences**.

XSS (Cross-Site Scripting)

Definition:
XSS is a **web security vulnerability** that allows attackers to inject malicious scripts into websites viewed by others.

Example:
If a website displays user input without cleaning it, an attacker might type:

```
<script>alert('Hacked!');</script>
```

Used in: Web security and ethical hacking.
Tip: Prevent XSS by **validating and escaping** all user inputs before displaying them.

XOR Gate (⊕)

Definition:
An **XOR gate** is a digital logic gate that outputs 1 (True) if the two inputs are **different**. If both inputs are the same, it outputs 0 (False).

Example (Logic Table):

A	B	Output
0	0	0
0	1	1
1	0	1
1	1	0

Used in: Electronics, AI circuits, data encryption.
Tip: XOR gates are the "brain" behind many bitwise and encryption algorithms.

XOR Operator (^) in Encryption

Definition:

In encryption, XOR is used to **combine plain text and key data**.
It produces a cipher text that can only be reversed with the same key.

Example:

```
a = ord('A')   # 65
b = ord('K')   # 75
print(chr(a ^ b))
```

Used in: Cryptography, data masking.
Tip: XOR encryption is simple but powerful when combined with secure keys.

XOR Swap Algorithm

Definition:

The **XOR swap** is a clever trick to swap two variable values without using a temporary variable.

Example:

```
a = 5
b = 7
a = a ^ b
b = a ^ b
a = a ^ b
print(a, b)   # Outputs: 7 5
```

Used in: Low-level programming, optimization.
Tip: Modern compilers optimize swapping automatically, but XOR swap shows how bit logic works!

LETTER Y

Y-Axis

Definition:

The **Y-axis** is the **vertical line** on a graph or coordinate system.
In programming, it represents **height** or **vertical position** — such as how far up or down something appears.

Example (Python – Matplotlib):

```
import matplotlib.pyplot as plt
plt.plot([1, 2, 3], [4, 5, 6])
plt.ylabel("Y-Axis")
plt.show()
```

Used in: Data visualization, game design, and coordinate-based graphics.
Tip: Think of the Y-axis as "up and down," while the X-axis goes "left and right."

YAML (YAML Ain't Markup Language)

Definition:
YAML is a simple data format used for **configuration files** and **data exchange**.
It uses indentation and symbols like : and - to organize data clearly.

Example (YAML file):

```
student:
  name: Raquel
  age: 7
  subjects:
    - Math
    - Science
```

Used in: Python, DevOps, Docker, APIs.
Tip: YAML is known for its readability — no brackets or braces, just clean indentation.

Yield (Python Keyword)

Definition:
The **yield** keyword is used inside a function to turn it into a **generator** — a function that can pause and resume, sending values one at a time.

Example:

```
def count_up():
    for i in range(3):
        yield i

for number in count_up():
    print(number)
```

Output:

```
0
1
2
```

Used in: Python.
Tip: Use `yield` when you need to process large amounts of data **one piece at a time** without using too much memory.

Yarn (JavaScript Package Manager)

Definition:
Yarn is a tool used to **install and manage packages** (pre-made code libraries) in JavaScript projects.
It's an alternative to npm and makes installation faster and more reliable.

Example (Command):

```
yarn add react
```

Used in: JavaScript, React, Node.js projects.
Tip: Yarn helps organize dependencies — think of it as a "librarian" for your web project's code.

Y-Coordinate

Definition:
The **Y-coordinate** defines an object's **vertical position** on a graph or screen.
It's often used with an **X-coordinate** to place items precisely.

Example:

```
position = (x, y)
print("Y-coordinate:", position[1])
```

Used in: Game development, graphics, robotics.
Tip: In most computer graphics, the Y-axis increases as you move **down**, not up.

Yarn.lock File

Definition:
The **yarn.lock** file keeps track of exact versions of packages installed in a project.
It ensures that everyone working on the same project has the same dependencies.

Used in: Web development with Yarn or npm.
Tip: Never delete `yarn.lock` — it guarantees your app runs the same everywhere.

Yes / No (Boolean Logic)

Definition:
In coding, **Yes** and **No** are represented as **True** and **False** — the foundation of **Boolean logic**.

Example:

```
is_logged_in = True
has_permission = False
```

Used in: All programming languages.
Tip: Computers make decisions using True/False logic — every condition, comparison, or test depends on it.

Y2K (Year 2000 Bug)

Definition:
The **Y2K bug** was a programming issue where older software used only two digits for the year (99 for 1999), causing problems when the year changed to 2000.

Example of the Problem:

```
99 + 1 = 00 (which systems read as 1900, not 2000)
```

Used in: Legacy systems and history of programming.
Tip: Y2K taught developers the importance of **date handling** and **data formatting.**

YUV Color Model

Definition:
The **YUV color model** separates an image into **luminance (Y)** and **chrominance (U and V)** — used for video compression and broadcasting.

Used in: Video codecs, digital TV, image processing.
Tip: YUV saves space while keeping colors accurate — it's how video streaming stays efficient.

Yield Statement (Advanced Use)

Definition:
A **yield statement** can send multiple values over time, pausing a function between outputs. It's ideal for loops, data streams, and infinite sequences.

Example:

```
def infinite_numbers():
    n = 1
    while True:
        yield n
        n += 1
```

Used in: Python generators, AI data pipelines.
Tip: Yield is like "pause and continue later" — very memory-efficient for long data tasks.

LETTER Z

Z-Axis

Definition:
The **Z-axis** represents **depth** in 3D space — how far something is in front of or behind another object.
It adds the third dimension to the X (width) and Y (height) axes.

Example (3D Graphics):

```
point = (x, y, z)
print("Depth:", point[2])
```

Used in: 3D graphics, simulations, robotics, and game design.
Tip: Think of Z as "forward and backward." It controls perspective and layering in 3D visuals.

Z-Index (CSS Property)

Definition:
The **z-index** in CSS determines which element appears **on top** when elements overlap on a webpage.
Higher numbers mean closer to the viewer.

Example (CSS):

```
.box1 {
  position: absolute;
  z-index: 2;
}

.box2 {
  position: absolute;
  z-index: 1;
}
```

Used in: Web design and front-end development.
Tip: Use z-index carefully — too many layers can cause layout confusion.

Zero (0)

Definition:
Zero represents "nothing" but has a powerful role in computing.
In logic, it means **False**; in arrays, it represents the **first position** (index 0).

Example:

```
numbers = [10, 20, 30]
print(numbers[0])   # Outputs: 10
```

Used in: All programming languages.
Tip: Remember: Computers count from **zero**, not one.

Zero-Based Indexing

Definition:
In programming, **zero-based indexing** means counting starts at 0 instead of 1.
So, the first element in a list has index 0, the second has index 1, and so on.

Example:

```
colors = ["red", "green", "blue"]
print(colors[0])  # Outputs: red
```

Used in: Python, C, Java, JavaScript, and most languages.
Tip: Forgetting zero-based indexing is one of the most common beginner mistakes.

ZeroDivisionError

Definition:
A **ZeroDivisionError** occurs when you try to divide a number by zero — something that's mathematically impossible.

Example:

```
x = 10 / 0  # ✖ Error!
```

Used in: Python and other languages.
Tip: Always check the denominator before dividing:

```
if y != 0:
    print(x / y)
```

Zip Function (Python)

Definition:
The **zip()** function combines multiple lists (or sequences) into pairs, making it easy to loop through related data.

Example:

```
names = ["Raquel", "Addison"]
ages = [7, 17]
for n, a in zip(names, ages):
    print(n, a)
```

Output:

Used in: Python.
Tip: Zip helps you merge data — like names and scores — neatly in loops.

Zombie Process

Definition:
A **zombie process** is a program that has finished running but still takes up space in memory because its parent process hasn't cleaned it up.

Used in: Operating systems, Linux, system programming.
Tip: Zombie processes don't use CPU, but too many can fill your process table.

Zoom (Scaling)

Definition:
Zoom refers to increasing or decreasing the size of an element, image, or view on screen. In web and graphics programming, it's used for dynamic resizing or visual effects.

Example (CSS):

```
div:hover {
  transform: scale(1.2);
}
```

Used in: Web design, image editing, and UI animations.
Tip: Always balance zoom effects with readability — too much scaling can distort layout.

Zone File

Definition:
A **zone file** is a text file that defines the mapping between **domain names** and **IP addresses** on the internet.

Example (DNS Zone Record):

```
example.com.  IN  A  192.0.2.1
```

Used in: Web hosting, DNS management, domain setup.
Tip: Editing zone files incorrectly can break a website — always double-check before saving.

Zero Padding

Definition:
Zero padding adds extra zeros to a number or string to maintain a uniform length.
It's often used in data formatting and binary operations.

Example:

```
print(str(7).zfill(3))  # Outputs: 007
```

Used in: Data formatting, encryption, digital signal processing.
Tip: Zero padding keeps data consistent for comparisons and sorting.

📖 Index

Epilogue —

Your Next Step Into the World of Code

When you opened this book, programming symbols might have felt like a secret language designed to keep you out. Brackets, operators, and punctuation marks looked intimidating — and tutorials often assumed you already understood them. But now, you've uncovered what they mean and why they exist. You've built the foundation many new coders miss.

You've learned that:

· Symbols are not random; they're the grammar and punctuation of code.

· Tiny differences — like = versus == — can completely change a program's meaning.

· Once you understand how symbols group, compare, assign, calculate, and connect, reading code becomes less about memorization and more about logical thinking.

This is where your real coding journey begins.

Understanding symbols won't make you a programmer overnight, but it removes one of the biggest early roadblocks. You now have the clarity to move into tutorials, courses, or projects with confidence. You can read code samples and understand what's happening instead of guessing. You can debug errors more easily because you recognize the role each symbol plays.

As you continue:

· Experiment. Play with the examples you've seen. Modify them. Break them on purpose — and then fix them.

· Build small things. A simple calculator, a guessing game, or an interactive web page can teach you more than theory ever will.

· Seek community. Join online forums, coding groups, or local meetups. Ask questions, share what you're learning, and help others who feel lost like you once did.

· Stay curious. Languages evolve, and so will your skills. Keep reading official documentation and exploring new tools.

Remember, every expert programmer started exactly where you are now — confused, curious, and trying to make sense of this new language. The difference is that you've taken the time to build a clear foundation instead of skipping the basics.

If this book helped you see code in a new way, you're already ahead of where I was when I started. You've unlocked the hidden language of programming, and that understanding will support everything you learn next — from your first print() statement to your first real application.

Keep learning. Keep experimenting. And whenever you feel stuck, remember: the symbols aren't there to confuse you — they're shortcuts created by programmers who were once beginners, too.

Your journey into coding doesn't end here.

It begins now — with clarity, confidence, and the power to keep going.

Happy coding,

Susie Hala